SPACES OF SECURITY AN.

Critical Geopolitics

Series Editors:
Klaus Dodds, Royal Holloway, University of London, UK
Merje Kuus, University of British Columbia, Canada

Over the last two decades, critical geopolitics has become a prominent field in human geography. It has developed to encompass topics associated with popular culture, everyday life, architecture and urban form as well as the more familiar issues of security, inter-national relations and global power projection. Critical geopolitics takes inspiration from studies of governmentality and biopolitics, gender and sexuality, political economy and development, postcolonialism and the study of emotion and affect. Methodologically, it continues to employ discourse analysis and is engaging with ethnography and participatory research methods. This rich field continues to develop new ways of analysing geopolitics.

This series provides an opportunity for early career researchers as well as established scholars to publish theoretically informed monographs and edited volumes that engage with critical geopolitics and related areas such as international relations theory and security studies. With an emphasis on accessible writing, the books in the series will appeal to wider audiences including journalists, policy communities and civil society organizations.

Spaces of Security and Insecurity
Geographies of the War on Terror

Edited by

ALAN INGRAM
University College London, UK

KLAUS DODDS
Royal Holloway University of London, UK

Routledge
Taylor & Francis Group

LONDON AND NEW YORK

First published 2009 by Ashgate Publishing

Published 2016 by Routledge
2 Park Square, Milton Park, Abingdon, Oxfordshire OX14 4RN
711 Third Avenue, New York, NY 10017, USA

First issued in paperback 2016

Routledge is an imprint of the Taylor & Francis Group, an informa business

British Library Cataloguing in Publication Data
Spaces of security and insecurity : geographies of the War
  on Terror
  1. Geopolitics 2. War on Terrorism, 2001- - Social aspects
  I. Ingram, Alan II. Dodds, Klaus
  320.1'2

Library of Congress Cataloging-in-Publication Data
Ingram, Alan.
  Spaces of security and insecurity : geographies of the War on Terror / by Alan Ingram
and Klaus Dodds.
      p. cm.
  Includes index.
  ISBN 978-0-7546-7349-1 -- ISBN 978-0-7546-9041-2 (e-book) 1. War on Terrorism,
2001- 2. Geopolitics. 3. Security, International. 4. National security. I. Dodds, Klaus. II.
Title.

  HV6431.I4775 2009
  909.83'1--dc22

                                                                    2008047392

  ISBN 13: 978-1-138-27058-9 (pbk)
  ISBN 13: 978-0-7546-7349-1 (hbk)

Contents

List of Figures and Plates

Figures

Plates

List of Contributors

Kezia Barker
Lecturer in Science and Environmental Studies, School of Continuing Education, Birkbeck, University of London.

Jason Dittmer
Lecturer in Geography, Department of Geography, University College London.

Klaus Dodds
Professor of Geopolitics, Department of Geography, Royal Holloway, University of London.

Stuart Elden
Professor of Political Geography, Department of Geography, Durham University.

Nick Gill
Lecturer in Geography, Department of Geography, Lancaster University.

Alan Ingram
Lecturer in Geography, Department of Geography, University College London.

Alex Jeffrey
Lecturer in Human Geography, School of Geography, Politics and Sociology, Newcastle University.

Lina Khatib
Lecturer in World Cinema, Department of Media Arts, Royal Holloway, University of London.

Merje Kuus
Assistant Professor, Department of Geography, University of British Columbia.

Timothy W. Luke
University Distinguished Professor of Political Science, College of Liberal Arts and Human Sciences, Virginia Polytechnic Institute and State University.

Nick Megoran
Lecturer in Human Geography, School of Geography, Politics and Sociology, Newcastle University.

Suthaharan Nadarajah
Lecturer in International Relations, Department of International Relations, University of Sussex.

Patricia Noxolo
Lecturer in Human Geography, Department of Geography, Loughborough University.

Richard Phillips
Reader in Geography, Department of Geography, University of Liverpool.

Chih Yuan Woon
Doctoral Researcher, Department of Geography, Royal Holloway, University of London.

Foreword

Space, to twist a well-known phrase, is not the final frontier, but instead the layers and lines of multiple spatialities reworked over and over again into the frontiers of some highly conflicted finality, like those international domestic crises discussed here in terms of security and insecurity. The inversions of meaning in today's war of terror both break and remake the security problematic of the Cold War, World War, and imperialism from those experienced during the anti-traditional, anti-fascist, and anti-communist times once dominated by the great powers. As the apparently once more definite divides between self/other, friend/foe, and ally/enemy fade into the ambiguities of the new times rising out of events in 2001, 1991 or 1941, the geographies of security and insecurity appear to be even more worthy topics of deliberate critical discussion.

Critical geopolitics is a most important and increasingly suggestive area of inquiry, unfolding at the conjuncture of social theory, political geography, international relations, and cultural studies. Anticipations of this intellectual project can be found in various works here and there during previous decades, but this approach to human geography clearly began to blossom in the late 1980s and early 1990s as the Cold War ground down to a stop. The nature and dominant discourses of security and insecurity – at the individual and group level – before and after 1989 change significantly in that watershed year, and critical geopolitics has become one of the key centers for investigating the characteristics of the post-Cold war world, as the analyses by the various contributors to this volume ably demonstrate. In turn, the shocks of 9/11 and the ensuing wars in Afghanistan and Iraq have further highlighted the importance of this area of research. These developments have provoked further conversations with other strands of inquiry, which are also reflected here in this volume.

To speak of security is to talk about insecurity. Without producing the latter, the need for the former is not obviously ever experienced, detected or considered. Hence, this volume makes a very significant contribution by remapping the geographies of terror off the conceptual coordinates of more articulated discourses focused on contested space, governance, exceptional authority, and illegitimacy. The desperate need in some states to envision the present essentially as the past restaged, renamed or revitalized is still quite tangible, but also often totally wrong-headed. The Bush administration's desperation to find an enemy equal to the USSR or Nazi Germany, for example, has brought forth the poorly understood 'war on terror' that is neither quite a life-and-death struggle with an 'Islamofascism' nor an effective response to the truly diverse terrorist groups – both foreign and domestic – threatening the world today.

Ingram and Dodds with their contributors rightly caution us to presume that 9.11.2001 was truly a decisive rupture in time, changing beyond a doubt the nature of security and insecurity forever. Such analytical presumption mistakes the geopolitical spin of the Bush, Blair and Brown governments, as the thickening vortices of failure, incompetence, and vainglory have gathered around them, for the solid substance of a new strategic paradigm. These authors look farther afield. In so doing, they turn up important new insights from critical geopolitics, postmodern international relations, feminist criticism, and postcolonial theory. The result is their much stronger sense of how to maintain continuity with worthy traditions in critical theory, political economy, and geopolitical analysis from the past, while testing important theoretical innovations in the present to meet the new security challenges of the all too immediate future.

Timothy W. Luke
Virginia Polytechnic Institute and State University

Preface: Placing the War on Terror

In September 2007 we organized a series of sessions at the annual conference of the Royal Geographical Society with the Institute of British Geographers in London. The intention was to take stock of the ways in which geographers had been advancing analyses of the US-led 'war on terror' and contemporary landscapes of security. While acutely aware that it has been just one of the stories shaping our world (albeit a peculiarly powerful one) the sessions considered how the war on terror has involved a series of geographical relationships and networks that have affected people and places in very different and unequal ways. The papers presented examined the effects of the war on terror and new dimensions of security and insecurity in a variety of spatial contexts, including Afghanistan, Iraq, Britain, the European Union, New Zealand, Philippines, Sri Lanka and the United States and within relations connecting these and other places. Most of them are reflected in this collection, along with the contributions of several other writers.

The sessions were framed in terms of critical geopolitics, but reached more widely into areas of geography and other disciplines with affinities to critical geopolitics more in spirit than letter, and were richer for it. We believe that while critical geopolitics has added much to the ways in which geographers and others analyse relationships between space and power, it must remain open to exactly this kind of cross-fertilization. This is not therefore a critical geopolitics collection *per se*, but one that seeks to counterpose different kinds of geographical thinking to the violence of geopolitics and security.

The chapters in this book warn against assuming that ideas about security and insecurity have changed beyond recognition. Rather, as a number of writers have observed, many developments that were visible before September 2001 have intensified and crystallized, sometimes in new ways, but invariably also reflecting the imprint of earlier spatializations.

The collection does not seek to add to the voluminous literature describing the foreign and security policies of the Bush and Blair administrations and their failings, but to connect with and extend the much smaller reflexive geographical literature examining (but also going beyond) these issues. The contributions to this volume consequently revisit some of the relatively well-known aspects of recent geopolitics, but with careful conceptual analysis that advances our knowledge of them still further. Others extend into terrains that are less often explored, such as the implications of current security policies for diasporic groups, links between action taken to secure mankind and action to secure ecosystems, and the responses of artists to contemporary geopolitics. Rather than make expansive, generalizing claims, they tend towards careful contextualization in terms of particular settings, and we believe that bringing this diverse set of explorations together adds depth as well as scope to our understanding of the geopolitical present.

We would like to thank chapter authors for their contributions to the book, and Peter Kennard and Cat Picton-Phillipps for permission to use their work on the front cover. We would also like to thank Val Rose, Sarah Horsley, Aimée Feenan, Bibi Stoute and Katy Low for their help and support in bringing the book together.

Finally, the book is dedicated to the geographers who supervised our doctoral research. At the University of Bristol, Klaus Dodds was fortunate to enjoy the constant encouragement and support of Professor Leslie Hepple who died unexpectedly in February 2007. He was a remarkable polymath equally at home with spatial econometrics as he was with studying geopolitics or the historical geographies of the north east of England. At the University of Cambridge, Alan Ingram was one of many students inspired by the research and teaching of Dr Graham Smith on geopolitics, nationalism and transformations in the Soviet Union and the post-Soviet states. His charismatic presence and incisive analysis have been sorely missed since his untimely death in April 1999.

Alan Ingram and Klaus Dodds
December 2008

Chapter 1

Spaces of Security and Insecurity: Geographies of the War on Terror

Alan Ingram and Klaus Dodds

Introduction

It is perhaps telling that the American television show, *24*, featuring the counter-terror operative Jack Bauer, is a big hit with defenders of the Bush administration's tactics since 11 September 2001. US Secretary for Homeland Security, Michael Chertoff, defended the show's controversial coverage of the use of torture as representing 'real-life' situations, and visited the set when it was shooting in Washington DC.[1] But the influence of the show has not been limited to the higher echelons of the Bush administration. Jack Bauer 'had many friends at Guantánamo', according to one US military lawyer who was stationed at the detention facility there: 'He gave people lots of ideas'.[2] Support also stretched to certain US lawmakers. When participants in a televised debate among potential Republican presidential candidates were asked to outline their position on torture during a period of presumed terrorist threat, Representative Tom Tancredo stated, 'I'm looking for Jack Bauer at that time, let me tell you'.[3] The office of the show's creator, Joel Surnow (himself well connected in conservative circles), is decorated by a US flag given to him by an army regiment during a visit to Baghdad. 'The military loves our show', he stated, 'People in the administration love the series too … It's a patriotic show. They *should* love it'.[4]

If James Bond was the iconic hero of the Cold War era, Jack Bauer (played by Kiefer Sutherland) occupies a central (and deeply problematic) spot in the popular US imagination of the war of terror. Over a number of series, Bauer's improbable exploits and willingness to use extreme measures are shown to be reasonable,

---

1   *New York Post*, 'Chertoff Meets Jack Bauer' (31 July 2008) at <http://www.nypost.com/seven/11082007/gossip/pagesix/chertoff_meets_jack_bauer_33039.htm>.

2   P. Sands, 'The Green Light', *Vanity Fair* (May 2008) at <http://www.vanityfair.com/politics/features/2008/05/guantanamo200805>.

3   R. Brooks, 'The GOP's Torture Enthusiasts', *LA Times* (18 May 2007) at <http://www.latimes.com/news/opinion/commentary/la-oe-brooks18may18,0,732795.column?coll=la-news-comment-opinions>.

4   J. Mayer, 'Whatever it Takes', *The New Yorker* (19 February 2007) at <http://www.newyorker.com/reporting/2007/02/19/070219fa_fact_mayer>. Emphasis added.

given the exceptional threats that are said to be facing America and the wider world. Yet while the show sometimes toys with ambiguity, it generally upholds a binary world view and a conventional mapping of homeland security culture. But the real enemy in *24* is less the terrorist than those who might call the veracity of this world view and security culture into question. Bauer's critics within the show may often be well-meaning, but they are always shown to be misguided. They are always proven by events to have chosen the wrong paradigm for dealing with variants on the 'ticking bomb' scenario, so favoured by defenders of torture. The show is less a reflection of US homeland security culture than a constitutive part of it.

Though 'homeland security' is the key referent in such discourses, as in earlier phases of geopolitical transformation, the globe itself has also re-emerged as the level at which the most important processes are assumed to operate and upon which security is presumed to rest. However, while there are many continuities with earlier colonial and Cold War eras, the conditions of contemporary globalization are enabling visions and strategies of security to become ever more expansive. One indication of this came in 2005 when Michael Chertoff spoke of an aspiration to create a 'world wide security envelope'. He asked, in the managerial argot adopted by certain kinds of security professionals,

> How do we move beyond simply partnering on an individual, episodic basis to building a true partnership that will operate in a mission–oriented focus where we will work together with our allies overseas to accomplish a mission that will secure the entire world?[5]

In Chertoff's vision, those inside the envelope would be treated with 'a high degree of confidence and trust'; however, this would be predicated upon 'the kind of in-depth analysis and the kind of in-depth vetting that is necessary to make sure those who seek to harm us do not slip through the cracks'.[6] Such visions, founded upon simplistic binaries of the kind often peddled by politicians and culture-makers alike, are likely to have profound geographical effects. This, then, is a maximalist vision of security that indicates how counter-terrorist operations in response to the attacks on the US of 11 September 2001 are linked into a much wider project: the goal of securing not just specific homelands but liberal globalization itself. This project is couched in terms of a simple imaginative geography, but it conceals a world of complexity. Numerous spaces of control have been created in many different places, connected by networks of communication, influence and coercion

---

5   As reported in *GovernmentExecutive.com*, 'Homeland Security Secretary Calls for "Worldwide Security Envelope"' (19 May 2005) at <http://www.govexec.com/dailyfed/0505/051905c1.htm>. The record of his comments 'as prepared' does not contain this exact phrase, but is consistent with it. See 'Remarks by Secretary of Homeland Security Michael Chertoff at the German Marshall Fund and the European Policy Centre' (23 May 2005) at <http://www.dhs.gov/xnews/speeches/speech_0253.shtm>.

6   Ibid.

that confound orthodox maps of world politics. These go beyond mere foreign interventions in places like Iraq and Afghanistan to produce complex topographies and topologies that are reframing the meaning of territory and sovereignty.[7] This book addresses the urgent task of documenting and interpreting these complex geographies and the ways in which people have contested them. It suggests that geographical imaginations are essential to any critique of the war on terror and emerging landscapes of security, and to the construction of alternatives.

Rather than reduce the complexity inherent in real geographies to simplistic periodizations and spatializations, it is also necessary to recognize the geopolitical present is constituted by multiple temporalities and multiple spatialities that exceed the states and security apparatuses, even as they are shaped by them.[8] In this context, attempts to identify moments or episodes as key or foundational can easily become partial or ethnocentric in ways that much recent geographical work has sought to unsettle. However, the 'war on terror' proclaimed by George W. Bush in 2001 has undoubtedly been an important catalyst of geographical transformation.

While the war on terror is not the main story for the majority of the world's people, the attempt to make it so represents an extraordinary attempt at geopolitical re-scripting and geographical remaking, the consequences and implications of which continue to reverberate. At the same time, it is necessary to be mindful of the fact that the term 'war on terror' does not denote a stable and coherent referent object, but is caught up in complex discursive struggles for legitimacy. One sign of this is the coincidence between British state actors moving away from the phrase itself, preferring instead to focus on supposedly more neutral ideas of criminality and radicalization, and its appearance as the premise for a satirical board game.[9] The term is also apt to reappear in other contexts as an ostensibly neutral descriptor, for example in the justification for increasing US military involvement in Africa, suggesting a certain durability and utility for some actors.[10] Thus, while most of the insecurities faced by most of the people in the world cannot and should not be connected to terrorism or counter-terrorism, this volume aims to trace ways in which the dominant visions of security associated with the war on terror have been woven into a variety of geographies and have in turn been called into question.

One key reference point for such an enterprise is the project of critical geopolitics. The idea of geopolitics as an active process by which the geographical complexity

---

7 S. Grey, *Ghost Plane: The True Story of the CIA's Rendition and Torture Program* (New York: St Martin's Griffin 2006); T. Paglen and A. Thompson, *Torture Taxi: On the Trail of the CIA's Rendition Flights* (Cambridge: Icon 2007).

8 D. Gregory, *The Colonial Present* (Oxford: Blackwell 2004).

9 On the former, see the chapter by Richard Phillips in this volume, note 8; on the latter <http://www.waronterrortheboardgame.com/>.

10 Statement of General William E. Ward, USA, Commander, United States Africa Command before the House Armed Services Committee (13 March 2008) at <http://armedservices.house.gov/pdfs/FC031308/Ward_Testimony031308.pdf>.

and richness of the world gets reduced to schematic spatial templates (such as East versus West or developed versus developing or international community versus 'axis of evil') has been around for two decades.[11] Critical geopolitics has not only explored how these templates are embedded within foreign policy and security discourses but also how they reverberate through popular culture and particular media such as film, newspapers, cartoons and television.[12] This basic insight has served as inspiration for the development of a large and growing body of scholarship that seeks to interrogate the spatialities associated with states and the inter-state system. This has contested reductive geopolitical visions and proposed a variety of techniques to re-conceptualize (and thereby perhaps remake) political geographies in less violent ways.

The ways in which critical geopolitics has addressed issues like these has shifted over time. Though they cannot be summarized in detail or with justice here, it is worth outlining some of the main orientations. First, there has been a recognition that struggles over space and power are not to be analysed only in terms of formal theories or processes of international relations; consideration is also required to examine the ways in which geopolitics circulates between formal theorizing, practical statecraft and popular domains. Second, recent developments, informed by renewed feminist critiques, have sought to bring attention to the material, embodied and emotional geographies around geopolitics in order to unsettle further the hierarchies and binaries upon which the practices of geopolitics are predicated.[13] Third, the empirical reach of critical geopolitics now takes in

---

11   J. Agnew, *Geopolitics: Re-Envisioning World Politics* (London: Routledge 1998); J. Agnew and S. Corbridge, 'The New Geopolitics: The Dynamics of Geopolitical Disorder', in R. Johnston and P. Taylor (eds), *A World in Crisis? Geographical Perspectives* (Oxford: Blackwell 1989) 266–88; J. Agnew and S. Corbridge, *Mastering Space: Hegemony, Territory and International Political Economy* (London: Routledge 1995); K. Dodds, *Global Geopolitics: A Critical Introduction* (Harlow: Prentice Hall 2005); G. Ó Tuathail, *Critical Geopolitics: The Politics of Writing Global Space* (London: Routledge 1996); G. Ó Tuathail and J. Agnew, 'Geopolitics and Discourse: Practical Geopolitical Reasoning in American Foreign Policy', *Political Geography* 11/2 (1992) 190–204; G. Ó Tuathail and S. Dalby (eds), *Rethinking Geopolitics* (London: Routledge 1998).

12   The literature on popular geopolitics continues to grow but includes, M. Power and A. Crampton (eds), *Cinema and Popular Geo-politics* (London: Routledge 2006) and J. Dittmer, 'The Tyranny of the Serial: Popular Geopolitics, the Nation and Comic Book Discourse', *Antipode* 39/2 (2007) 247–68. One of the earliest contributions was J. Sharp, 'Hegemony, Popular Culture and Geopolitics: The Reader's Digest and the Construction of Danger', *Political Geography* 15/6–7 (1996) 557–70.

13   In particular R. Pain, 'Globalized Fear? Towards an Emotional Geopolitics', forthcoming in *Progress in Human Geography*; R. Pain and S. Smith (eds), *Fear: Critical Geopolitics and Everyday Life* (Aldershot: Ashgate 2008). Also J. Hyndman, 'Beyond Either/Or: A Feminist Analysis of September 11th', *ACME: An International E-journal for Critical Geographies* 2/1 (2003) 1–13; J. Hyndman 'Mind the Gap: Bridging Feminist and Political Geography Through Geopolitics', *Political Geography* 23/3 (2004) 307–22;

everything including the spatial patterning of violence; maps, speeches, policy documents and media reports; film (e.g. *The Kingdom* 2007), cartoons (e.g. those of Steve Bell), radio (e.g. Voice of America), video games (e.g. America's Army) and pop songs (e.g. Angry American). It is also engaging much more fully with the ways in which people experience, respond and contest geopolitics and security cultures.

This broad project has emerged and evolved in affiliation with other currents in geography, the social sciences and humanities. In particular, there have been close relationships with parts of international relations and critical security studies.[14] But there have also been engagements with cultural studies and with social geography, with urban geography, and with the literature on imaginative geographies, with which critical geopolitics shares many intellectual roots.[15] While there have been tensions and debates between different parts of this intellectual landscape, the call to regard domains often considered separately (such as 'international relations' and 'everyday life') as 'part of the same assemblage' seems to be an increasingly important theme for work in the field.[16] Identifying connections and renewing ethical and political concern for the ways in which people contest, transcend or remake current dividing lines (reminiscent of calls by geographers more than a century ago) is another.[17]

The contributions to the book engage with this constellation of concerns in different ways. They draw on a range of theoretical inspirations (including

J. Sharp, 'Remasculinizing Geopolitics? Comments on Gerard Ó Tuathail's Critical Geopolitics', *Political Geography* 19/3 (2000) 361–4. For a consideration of the relevance of sexuality to the war on terror, see J. Puar and A. Rai 'Monster, Terrorist, and Fag: The War on Terrorism and the Production of Docile Patriots', *Social Text* 20/3 (2002) 117–48.

14   See, for example, R. Ashley, 'The Geopolitics of Geopolitical Space: Toward a Critical Social Theory of International Politics', *Alternatives* 12/4 (1987) 403–34; K. Booth (ed.), *Critical Security Studies and World Politics* (London: Lynne Rienner 2005); B. Buzan, O. Waever and J. de Wilde, *Security: A New Framework for Analysis* (London: Lynne Reiner 1998); D. Campbell, *Writing Security* (Manchester: Manchester University Press 1998); J. Der Derian and M. Shapiro, *International/Intertextual Relations: Postmodern Readings of World Politics* (Lexington: Lexington Books 1989); C. Enloe, *Bananas, Beaches and Bases: Making Feminist Sense of International Politics* (Berkeley: University of California Press 1990); K. Krause and M. Williams (eds), *Critical Security Studies* (London: UCL Press 1997).

15   Notable recent contributions include D. Cowen and E. Gilbert (eds), *War, Citizenship, Territory* (London: Routledge 2007); S. Graham (ed.), *Cities, War and Terrorism: Towards an Urban Geopolitics* (Oxford: Blackwell 2004); D. Gregory, *The Colonial Present* (note 8); D. Gregory and A. Pred (eds), *Violent Geographies: Fear, Terror and Political Violence* (London: Routledge 2007); P. Rajaram and C. Gundy-Warr (eds), *Borderscapes* (Minnesota: University of Minnesota Press 2007); G. Ó Tuathail, S. Dalby and P. Routledge (eds), *The Geopolitics Reader*, second edition (London: Routledge 2006).

16   Pain and Smith (note 13) p. 3.

17   P. Kropotkin, 'What Geography Ought to Be', *The Nineteenth Century* 18 (1885) 940–56.

imaginative geographies and Foucauldian governmentality as well as critical geopolitics), methodological approaches (including discourse analysis, interviewing, online research and ethnography) and empirical materials (ranging from policy documents to art works). While all of the contributions are consistent with the idea that geopolitics and security are better considered as forms of situated knowledge rather than abstract truths about the world, we recognize too that this book in general reflects its production from within the UK by authors who, while their life histories may trace diverse geographies, are mostly (though not exclusively) based there.[18]

To place the focus of our critique on security and the war on terror, is not to deny that threats exist. However, we believe that it is important to bring practices of security much more fully into view and to examine their implications. One reason for this is that many security practices tend to protect the richer and more powerful and to target the already-excluded. Another concern is that the rush to secure existing forms of social relations risks leaving the deep causes of violence and human vulnerability largely untouched. So security practices themselves must be subject to critical scrutiny as part of a wider study of socio-spatial dynamics. To focus exclusively on 'threats' and to screen out 'security' as a constitutive factor in the political geographies now unfolding (as some would prefer) is to render analysis partial and to play into the hands of the many actors who have consciously performed security discourse in order to promote their own interests and diminish their own accountability.[19]

Connecting Themes

While each of the chapters in the book represents a distinctive contribution, there are a number of connecting themes that link them together.

One major theme concerns the spatial vocabulary of international relations.[20] Terms such as 'homeland', 'international community', 'failed state', 'terrorist network' and 'rogue state' direct the gaze towards a particular version of the present as already-given and unproblematic. However, it is necessary to historicize our

---

18   K. Dodds and D. Atkinson (eds), *Geopolitical Traditions: A Century of Geopolitical Thought* (London: Routledge 2000).

19   See the discussion of parapolitics in P. Scott, *Drugs, Oil and War: The United States in Afghanistan, Colombia and Indochina* (Lanham: Rowman & Littlefield 2003) and P. Scott, *The Road to 9/11: Wealth, Empire and the Future of America* (Berkeley: University of California Press 2007); see also J. Butler *Precarious Life* (London: Verso 2004).

20   J. Der Derian, 'Decoding the National Security Strategy of the United States of America', *Boundary* 2/30 (2003) 19–27; A. Kaplan 'Homeland Insecurities: Reflections on Language and Space', *Radical History Review* 85 (2003) 82–93; A. Simons and D. Tucker, 'The Misleading Problem of Failed States: A "Socio-Geography" of Terrorism in the Post-9/11 Era', *Third World Quarterly* 28/2 (2007) 387–401; J. Sidaway, 'The Dissemination of Banal Geopolitics: Webs of Extremism and Insecurity', *Antipode* 40/1 (2008) 2–8.

understanding of the geographies that they help to constitute. The war on terror and other security projects are unfolding not across some kind of geopolitical *tabula rasa* but across landscapes that bear the marks of pre-existing and multiple struggles over enduring colonial, Cold War and other geopolitical orders.

In particular, the brunt of the spatial reordering associated with the war on terror has been borne by people in and from the global South, especially countries such as Afghanistan, Iraq and neighbouring countries such as Pakistan.[21] There has been renewed concern among western security policymakers with 'ungoverned' spaces (for example in Pakistan's tribal regions, or around the Sahara), in some ways recapitulating concerns from the 1990s about the spaces of criminal and drug economies.[22] And cities (often divided according to a series of binary types corresponding to north/south) have increasingly become the target of geopolitical strategies.[23] Underdevelopment has been linked to anxieties about zones of instability, failed cities and failed states, as western governments worry that such regions might 'export' security threats, or that regions where strategic resources are located might be taken out of the global economy.[24] International development is thus called to perform functions that are bio-political as well as geopolitical: to fortify fragile states and stabilize civilian populations in global danger zones; stem refugee and migrant flows from them; and protect core states' welfare systems from undesirables.[25] Security now resides alongside competent neoliberal economic management as a marker of 'good governance'.[26]

21　C. Bhatt, 'Frontlines and Interstices in the Global War on Terror', *Development and Change* 38/6 (2007) 1073–93; N. Ethlinger and F. Bosco, 'Thinking Through Networks and their Spatiality: A Critique of the US (Public) War on Terrorism and its Geographical Discourse' *Antipode* 36/2 (2004) 249–71. On colonial and post-1945 spatial orderings of geopolitics see D. Slater, *Geopolitics and the Post-Colonial* (Oxford: Blackwell 2004).

22　M. Castells 'The Perverse Connection: The Global Criminal Economy', in *End of Millennium* (Oxford: Blackwell 1998) 165–203; J. Mittelman and R. Johnston 'The Globalization of Organized Crime, the Courtesan State and the Corruption of Civil Society', *Global Governance* 5/1 (1999) 103–26; A. Rabasa, S. Boraz, P. Chalk, K. Cragin, T. Karasik, J. Moroney, K. O'Brien and J. Peters, *Ungoverned Territories: Understanding and Reducing Terrorism Risks* (Santa Monica: RAND 2007).

23　Graham (note 15).

24　See, for instance, A. Ingram, 'HIV/AIDS, Security and the Geopolitics of US–Nigerian Relations', *Review of International Political Economy* 14/3 (2007) 510–34.

25　M. Duffield 'Getting Savages to Fight Barbarians: Development, Security and the Colonial Present', *Conflict, Development and Security* 5/2 (2005) 141–59.

26　M. Dillon, 'Governing Through Contingency: The Security of Biopolitical Governance', *Political Geography* 26/1 (2007) 41–67; D. Pease, 'The Global Homeland State: Bush's Bio-Political Settlement', *Boundary* 2/30 (2003) 1–18. See also the useful paper by S. Weber, 'War, Terrorism and Spectacle: On Towers and Caves', *The South Atlantic Quarterly* 101/3 (2002) 449–58.

A second set of issues arises from the ways in which ideas of security are used to invoke a special kind of politics, involving exceptional prerogatives, emergency measures, recourse to violence and the reassertion of sovereignty to counter threats to the body politic.[27] Securitization theory, Foucault's idea of governmentality, and Agamben's writings on the state of exception have been particularly influential in recent reflexive analyses of security, and they are signalled in different ways in the chapters of this book.[28] Each of these perspectives has produced insights into the ways in which the powerful can manipulate institutions and ideas in the service of their interests. They have also revealed the inherently problematic nature of such moves and the need to consider 'security' as part of politics and not outside it. Developing this theme, geographical work has called for a consideration of space as part of any conceptualization or indeed critique of security. For example, geographers have sought to clarify the distinctly geographical ways in which 'exceptional' spaces are produced; they are hardly the non-places they have often been claimed to be, and recognizing this is an important resource for repoliticizing security.[29] What we wish to draw attention to here then are the ways that security takes place and is played out across and through space, via feedback, interplay and mutual constitution between violence, the politics of security and diverse landscapes. These range from Afghanistan, Iraq and extraterritorial prisons, to the control of asylum and immigration, to the spaces of diaspora, of art and even of gardening in places apparently far removed from the war on terror such as Christchurch, New Zealand.

A third thread is concerned with the governmentality of borders and mobility. Geographers have long been aware of the complexity of borders as places where distinctions between inside and outside, domestic and foreign, have been mediated. However, border practices are taking on new forms and new significance in response to (often rather ethnocentric) fears that globalization is bringing about a collapse between domestic and foreign spheres. As a result, border practices are no longer confined to the physical boundaries of states but are being disseminated across 'domestic' and 'foreign' spaces. And as the introductory quote from Michael

---

27   M. Dean, *Governing Societies* (Maidenhead: Open University Press 2007).

28   On securitization see Buzan et al., *Security: A New Framework for Analysis* (note 14); G. Agamben, *Homo Sacer* (Stanford: Stanford University Press 1998) and *State of Exception* (Chicago: University of Chicago Press 2005); on governmentality and security see in particular D. Bigo, 'Security and Immigration: Towards a Critique of the Governmentality of Unease', *Alternatives* 27/1 (2002) 63–92; Dean (note 27).

29   For geographical engagements with Agamben see C. Minca, 'The Return of the Camp', *Progress in Human Geography* 29/4 (2004) 405–12; C. Minca, 'Agamben's Geographies of Modernity', *Political Geography* 26/1 (2007) 78–97; D. Gregory, *The Colonial Present* (note 8); D. Gregory, 'Vanishing Points', in Gregory and Pred (note 15) 205–36; D. Gregory, 'The Black Flag: Guantánamo Bay and the Space of Exception', *Geografisker Annaler, Series B: Human Geography* 88/4 (2006) 405–27.

Chertoff suggests, they are also being called upon to filter undesirables out of global flows in 'smart' ways.[30]

While anxieties about the figures of the immigrant and the asylum seeker have provoked a reinforcement of border practices and their dissemination throughout the spaces of the state and inter-state system, other flows of money, professional personnel and goods are treated differently.[31] Taken together, this emerging regime of inclusion and exclusion is proving instrumental in helping to secure particular identity claims in the United States, Britain and across the EU, while producing a network of surveillance, detention and deportation with material and embodied consequences. Gargi Battacharrya has expressed the core contradictions of anxieties about mobile populations:

> What is increasingly apparent is the ... gap between public rhetoric and practical engagement – the first screams horror at the dangerous consequences of global integration, whereas the second happily exploits the inequalities that the global economy throws up.[32]

As Battacharrya implies, from a certain perspective such contradictions are deeply functional, producing marginalized populations available for untrammelled exploitation. One influential reference point for the analysis of the securitization of immigration and asylum has been Didier Bigo's discussion of the 'governmentality of unease'.[33] Bigo suggests that a wide spectrum of actors have promoted security as a prism through which to view and govern diverse phenomena.[34] This is less to do with the intrinsic nature of those phenomena or attempts to defend identity, he suggests, than a way of promoting particular bureaucratic, political and professional interests and of covering up failures of government:

> The popularity of this security prism is not an expression of traditional responses to a rise of insecurity, crime, terrorism, and the negative effects of globalization; it is the result of the creation of a continuum of threats and general unease in which many different actors exchange their fears and beliefs in the process of making a risky and dangerous society. The professionals in charge of the management of

---

30   See for example P. Rajaram and C. Gundy-Warr, 'Introduction', in Rajaram and Gundy-Warr (note 15) ix–xl; M. Salter, 'Passports, Mobility and Security: How Smart Can the Border Be?', *International Studies Perspectives* 5/1 (2004) 71–91.

31   M. Coleman, 'A Geopolitics of Engagement: Neoliberalism, the War on Terrorism, and the Reconfiguration of US Immigration Enforcement', *Geopolitics* 12/4 (2007) 607–34.

32   G. Battacharrya, *Traffick: The Illicit Movement of People and Things* (London: Zed 2005) p. 156.

33   Bigo (note 28).

34   See also M. Hannah 'Torture and the Ticking Bomb: The War on Terror as a Geographical Imagination of Power/Knowledge', *Annals of the Association of American Geographers* 96 (2006) 622–40.

> risk and fear especially transfer the legitimacy they gain from struggles against terrorists, criminals, spies, and counterfeiters toward other targets, most notably transnational political activists, people crossing borders, or people born in the country but with foreign parents.[35]

Though he does not explore the effects of this governmentality in great empirical detail, those caught up in what Bigo calls the security continuum are subsequently liable to suffer exclusion and domination in a variety of ways, ranging from refusal of access to welfare to incarceration and torture.[36] At the same time, in some places (most notably the US), unease and security concerns are woven through the fabric of consumer society and domesticity, as citizens have been encouraged to secure their bodies, domestic environments and children against terrorist and other threats.[37] Yet while forms of parental/communal hypervigilance have been encouraged, public debate about other conceptions or dimensions of security deriving from, for example, ideas of social justice, have been curtailed. The exclusion of those on the margins of citizenship is thus compounded in the name of protecting a particular vision of the homeland.[38]

While noting the force of such technologies of power, geographers have become concerned about expansive or epochal claims about fear and security, and have sought to contextualize and evaluate their effects within specific settings.[39] A number of the contributions to this volume are similarly concerned with the ways in which governmentality literally takes place, at specific sites and through scalar processes, and particularly in relation to mobile or transnational populations or entities.[40]

Fourth, security is often about the routine as well as the exceptional. While the kind of vetting and investigation mentioned by Michael Chertoff is increasingly undertaken using sophisticated technology, it also relies on workers to administer the system and execute daily judgements. Security concerns enter into the business of NGOs formerly concerned with development, asylum and immigration, or humanitarian work. It is not just liberty and citizenship that become entwined with security, but charity too. Members of groups facing routine suspicion have to learn how to negotiate the homeland security state. As transport networks have become a focus for terrorist attacks, they become places where certain kinds of

---

35   Bigo (note 28) p. 63.

36   For an anthropological account of 'outcast life', see M. Agier, *On the Margins of the World* (Cambridge: Polity 2008).

37   C. Katz, 'Me and My Monkey: What's Hiding in the Security State', in Pain and Smith (note 13) pp. 59–74.

38   W. Walters, 'Secure Borders, Safe Haven, Domopolitics', *Citizenship Studies* 8/3 (2004) 237–60.

39   See Pain; Pain and Smith (note 13).

40   J. Crampton and S. Elden (eds), *Space, Knowledge and Power: Foucault and Geography* (Aldershot: Ashgate 2007).

people are criminalized by implication. All of this produces everyday geographies that form the flip side of the 'new normal' and its institutionalization of exceptional legal provisions and surveillance. At the same time, these everyday geographies of security, just as much as the headline events of the war on terror, become a terrain of opposition and critique.

The fifth theme, then, relates to the inherently contested nature of the geographies at the heart of the war on terror and landscapes of security. While human geographies are also geometries of power, these geometries have to be continually remade, and people invariably find ways to reimagine and contest them, albeit in different ways and to different extents. Creativity, ingenuity, hope, care and disrespect often coexist with a blanket refusal to stick to the scripts handed down by power holders. While recognizing the force that the powerful can exert at particular geopolitical moments, care is required that analytical frameworks do not reinforce the power geometries and hierarchies they seek to disrupt, thereby writing certain kinds of people and places out of the story and overstating the influence of the powerful.[41]

So while we are mindful of the globalizing nature of the war on terror and related security projects, we are also unwilling to assume that people and places are affected equally or unable or unwilling to contest things. Indeed, opposition and alternative visions have been manifest in a whole variety of ways. It is clear that a host of individuals, organizations and networks, from antiwar movements to the US Supreme Court, continue to play an important role in contesting the cultural, political and legal geographies of the war on terror. Yet at the same time, the nature of contemporary neoliberal globalization, governmentality and security means that an increasingly wide circle of people and places are drawn into power relations (as workers, as passport holders, as tax payers, as consumers, as subjects and as bodies) in ways that are difficult to escape or control. As Arundhati Roy has said of the reach of globalization:

> There's no innocence and there isn't any sense in which any of us is perfect or not invested in the system ... One is not powerful enough nor powerless enough not to be invested in the process. Most of us are completely enmeshed in the way the world works. All our hands are dirty.[42]

This poses challenges for ideas of democracy, dissent and resistance, especially as some see the war on terror as facilitating moves towards their preferred state of affairs. For example, certain evangelical Christian groups read the war on terror as part of a divine plan for the annihilation of non-believers and the eventual arrival of a new kingdom of Christ. As Hoff has noted, like geopolitical disaster, 'environmental destruction (and even the frightening rise in personal and national debt) is fatalistically welcomed by the vast majority of believers in the Rapture

---

41   Pain and Smith (note 13).

42   A. Roy, *The Chequebook and the Cruise Missile* (London: Harper Perennial 2004) p. 32.

as a sign of the anticipated apocalypse'.[43] While attentive to the possibilities for resistance and critique and the expansion of geographies of responsibility and care, accounting for the diversity of cultures through which geopolitics and security are produced and renegotiated is also a crucial task of analysis.

The Contributions

While each of the above themes run through the contributions in different ways, we have structured them loosely in terms of three broad concerns. The first part (Constructing the War on Terror), explores some of the ways in which the war on terror has been established geographically, considering the shifting and often incoherent permutations of sovereignty and international law (Elden), the designation of states as rogue, failed or fragile (Jeffrey), the imaginative geographies of commemoration (Megoran) and the declaration by the US of a 'second front' to the war on terror in South East Asia (Woon). These contributions illustrate the centrality of spatiality to the conduct of the war on terror and show how the powerful have (intentionally but also inadvertently) manipulated space and perpetuated state-centric imaginations of political community, even as they have exploited the interstices of the political map and prised open new spaces in their attempts to control people and places. These contributions also offer incisive critiques of these processes and propose alternative readings and approaches.

Stuart Elden's chapter considers the role of former British Prime Minister Tony Blair in the war on terror by examining his actions in the run up to the invasion of Iraq in the context of wider debates about territorial integrity and humanitarian intervention. From the late 1990s, Blair was an advocate of what he called the doctrine of international community and proposed that the latter should intervene in response to humanitarian emergencies engendered by conflict and authoritarianism. At the same time, neoconservative intellectuals in the US were elaborating arguments that established notions of non-intervention and equal sovereignty of states would need to be qualified in the face of new levels of threat. In his detailed analysis of the events leading up to the 2003 invasion of Iraq, Elden shows how the territorial integrity of that country was preserved at the very moment when it was about to be violated, with or without United Nations approval. For Blair, a form of territorial preservation was a *sine qua non* of stability and prosperity not only in Iraq but also elsewhere in the region. On the other hand, Blair found himself involved with a US-led project that was fundamentally designed to impose a conditional form of sovereignty on Iraq itself. Elden's account provides particular insights into how concepts that are supposedly foundational to global order (such as sovereignty) or meant to be about the protection of people caught up

---

43   J. Hoff, *A Faustian Foreign Policy* (Cambridge: Cambridge University Press 2008) p. 191.

in violence and repression (such as humanitarian intervention) can be manipulated and rendered provisional in the context of geopolitical struggles.

Alex Jeffrey's chapter develops these concerns by considering the designation of states considered problematic for world order in terms of particular labels ('rogue', 'failed' and 'fragile') that have come to the fore in global security discourse. Rather than follow certain academic and policy analysts in taking such labels as neutral descriptions of policy problems amenable to technical interventions, Jeffrey explores their productive character and relates them to key dimensions of global power relations and their contradictions, in terms of the embedded statism of contemporary geopolitics. Jeffrey's analysis considers how these terms have been applied and the consequences they have had not only for our understanding of sovereignty and statehood but also for people and communities found within territories thus designated. While people and places are caught up in transnational networks of power, the embedded statism of the 'international community' tends to locate causality and responsibility within the nation-state. Focusing particularly on the case of Iraq, Jeffrey demonstrates how discourses associated with these concepts were used to justify not only intervention but also subsequent attempts to re-make the country in the name of neoliberalism, state competence and global security. As this project has foundered, the occupying powers have sought to disavow their own responsibility and project it onto the nascent Iraqi state and regional actors.

Nick Megoran shows in his chapter that the commemorative practices associated with the attacks on the US of 11 September 2001 deserve careful consideration if we are to understand the ways in which subsequent projects have been legitimized. Focusing on the service of remembrance held at St Paul's Cathedral in London on 14 September 2001, he considers how expressions of grief and sorrow were planned, managed and performed. Drawing on interviews with the planners of the service, Megoran shows how they tried to depoliticize it in order to avoid giving any kind of offence to the mourners. However, while the actors involved sought to speak to the shock and pain felt by many people, Megoran shows how commemoration was already being shaped by geopolitical logics. For example, eschewing earlier, more complex positions within the Church of England on questions of militarism, the Archbishop of Canterbury's sermon framed the attacks as an evil act inflicted upon the freedom and security of America, and aspects of the service contained distinctly militaristic overtones. Megoran (drawing on Judith Butler) argues that the service both inserted the United States at the head of a hierarchy of grief and played into the momentum for militarized responses.

Chih Yuan Woon's chapter draws on the work of Derek Gregory to interrogate the ways in which South East Asia has been caught up in the war on terror through an examination of the imaginative geographies present in ideas (such as the 'Second Front') used by US policymakers to frame events in the region and their response to them. He focuses in particular on the Philippines and the island of Mindanao, described by the acting US Ambassador as a 'new Mecca for terrorism'. As Woon notes, however, such simplistic characterizations of place and reductive ideas about Islam and politics in the region obscure much of its complex history and geography, which have also been shaped by its societies' encounters with expansionary Christianity, Spanish and American colonialism and Cold War geopolitics. Increasing US-Philippine security cooperation has been premised on the idea of countering 'global networks' of 'Islamic terrorism', but Woon shows how separatist struggles such as that of the Moro National Liberation Front cannot be understood in terms of the explanatory categories of the war on terror. He argues that the rhetoric of the 'second front' is ultimately self-defeating and calls for more cosmopolitan geographical imaginations that are sensitive to the multiple connections, affiliations and interdependencies between people and places.

The second part (Governing through Security) traces some of the wider ripple effects of the war on terror and contemporary security practices. These chapters show the influence of the imaginative geographies of security and the war on terror and how they are negotiated at specific sites. They consider: the effects of the designation of the Liberation Tigers of Tamil Eelam (LTTE) as a terrorist organization on the Tamil diaspora (Nadarajah); the responses of UK NGOs concerned with asylum and immigration (Noxolo); the UK asylum and immigration system (Gill); the construction of New Zealand's biosecurity regime (Barker); and the resonance between talk about democracy and military 'capabilities' in the EU and NATO (Kuus).

Suthaharan Nadarajah considers the implications of the international counter-terrorism regime for the Tamil diaspora living in London and Canada. Drawing on critiques of liberal peace, he explores the tension between geopolitical architectures and ideas of democracy, human rights and the principle of self-determination. He argues that the designation of the LTTE has in effect criminalized political resistance to state repression and subjected the activities of many Tamil people and organizations in the diaspora to a powerful disciplinary framework. Alongside ever more strict immigration and asylum restrictions, this has had profound implications for the ability of Tamil expatriates to engage in any form of activity (including charitable projects) that might be construed as threatening to the territorial integrity of the Sri Lankan state. The only response that will be considered legitimate under the counter-terrorist regime is recourse to liberal norms and processes that many Tamils believe cannot operate meaningfully within existing geopolitical structures. However, as Nadarajah shows, such disciplinary strategies are failing and have in fact fostered repoliticization within the Tamil diaspora.

Drawing in part on Bigo's account of the governmentality of unease, Patricia Noxolo investigates the strategic responses of a number of British non-governmental organizations to the securitization of immigration and asylum. Noxolo's concerns lie with how a variety of groups have responded to the war on terror through multi-scalar networking and discursive repositioning. Analysing textual and interview materials, the chapter explores the difficulties that are faced by groups seeking to challenge a prevailing defensive localism which often clouds distinctions between the immigrants, asylum seekers and terrorists. In particular, the chapter draws out the difficulties of challenging the idea of intrinsic connections between immigration and insecurity without reinforcing security discourse itself. She emphasizes the point that the link between immigration and insecurity is not natural or pre-given and that considerable reflexivity is required on the part of all kinds of actors if this link is not to be reinforced in the UK and elsewhere.

Nick Gill considers the ways such ideas are propagated, contested and negotiated in and around Lunar House. This is a nondescript but important building located in London, where many people make claims for refugee status. It is also adjacent to the headquarters of the Borders and Immigration Agency (formerly the Immigration and Nationality Directorate). Lunar House has been heavily caught up in the securitization of asylum and immigration in the UK, but rather than focus primarily on press coverage, or statements by politicians and security professionals, or the experience of people seeking asylum, Gill focuses on the experiences of the people who work in the building, either running the asylum process or supporting people through it. Drawing on interviewing and ethnographic research, he considers how the fear, mistrust and suspicion associated with the war on terror and security discourse play out in the mundane, banal and everyday bureaucratic spaces of immigration control. The chapter shows how anxieties created by a 24/7 media reporting culture become embedded within bureaucratic rationales, the spatial organization of the building and the administrative routines of staff, with a series of negative and self-reinforcing consequences. He suggests in conclusion that the circulation of unease at Lunar House parallels processes at work in other places.

Kezia Baker's chapter examines the emergence and implications of associations between anthropo-security and biosecurity in New Zealand. As a particular kind of postcolonial settler state, New Zealand has had a troubled relationship with ideas of civilization and cross-border mobility, and the chapter considers how these have been worked out in the construction of a biosecurity regime designed to protect 'native' plant species from 'alien' invasion. Barker shows how biosecurity policy makers and practitioners have appealed to images of war, security and terrorism in order to incite support and compliance among citizens for measures which may force them to eliminate plant species to which they feel attached. At the same time, drawing on ethnographic research at garden shows and interviews with biosecurity managers, she elucidates the ambivalence and dilemmas produced by reliance on discourses that depend upon a denigrated other, and whose ethical status is problematic in the context of New Zealand's own anxious engagement with postcolonialism. Though Barker is sympathetic towards the predicament of biosecurity managers, she draws out the dangers inherent in any appeal to security discourse, regardless of the motivations or goals of the people involved.

Merje Kuus investigates discourses of security around NATO and the EU. Created in 1949, NATO has proven more resilient than many predicted. Like the EU, its membership has grown since the ending of the Cold War. Kuus considers how discussions within NATO and the EU about Europe's 'neighbourhoods' have been caught up in dynamics of militarization and securitization. In particular, she considers how the strategies of security policy makers and bureaucrats are normalized and depoliticized through 'capabilities talk'. Focusing specifically on Estonia, she shows how geopolitical discourse has shifted in states that have gone from being 'consumers' of security to 'providers'; new and aspiring EU member states can now reinforce their place in 'the West' by assisting in global security projects beyond its limits. Drawing on the work of William Connolly, Kuus shows how militarism is normalized and an ever-more pervasive security agenda is propagated via association with a series of benign and de-politicized concepts in a 'resonance machine' that is resistant to analysis. She suggests that although not in explicit alliance, the synchronization of capabilities talk between the EU and NATO is helping to ensure the continuation of militarization within the European continent and beyond.

The final part (Alternative Imaginations) considers more specifically how different people in different places have actively resisted, negotiated or embraced the war on terror. It considers the role of satellite television in transforming the broadcasting and security landscapes of the Middle East (Khatib); an emerging geopolitical culture associated with a particular kind of American evangelical Christianity that welcomes the insecurity associated with the war on terror (Dittmer); the geographical imaginations and anti-imperialism of British antiwar groups (Phillips); and the ways in which artists from a variety of places, coming together in the Hague, have critiqued the imaginative geographies of the war on terror and proposed their own counter-geographies (Ingram).

The changing regional geographies of satellite television in the Middle East are the subject matter of Lina Khatib's chapter. She shows how the launch of *Al-Jazeera* in 1996 had a major impact on the broadcasting landscapes of the region and within diaspora communities in Europe and North America. In the aftermath of the attacks on the US of 11 September 2001, satellite television has grown in importance within the Arab world as more channels and networks have been established by competing elites. The visual and political-economic landscape around satellite media has been heavily influenced by territorial struggles such as those involving Israel-Palestine, Israel-Lebanon-Syria and Israel-Iran alongside the deeply controversial and resented US invasion of Iraq in 2003. Khatib shows how local satellite television stations became important but problematic elements in the representational politics surrounding the 2006 Lebanon war and broader attempts to democratize the public sphere within the Arab world. With the region also experiencing renewed attempts by American, British and Russian television stations and their political sponsors to influence public opinion, she is not hopeful that dialogue and peaceful exchange will result.

In his chapter Jason Dittmer explores a geopolitical culture that challenges liberal assumptions about the meanings of security and insecurity. Drawing on online research, he discusses how contributors to a web forum for premillennial dispensationalists are interpreting geopolitical events in terms of their religious beliefs and relating them to their own emotional security. Premillennial dispensationalism forms a significant stream within American evangelical Christianity, and has been credited with helping to secure a two-term Presidency for George W Bush. Its adherents live in expectation of an apocalyptic series of events that will culminate in the Rapture, when believers will ascend to heaven with Jesus and non-believers will be left behind. Dittmer shows how believers posting to a particular web forum both welcome violence and insecurity in the Middle East as portents of the Rapture, and evaluate their relationships with others in light of what they believe will happen as it approaches. He also traces the disciplinary mechanisms at work in the forum. Though they are pessimistic with regard to the possibility of peace, participants often find relief from their anxiety in contemplation of their place in a perfectly just kingdom of Christ that will last a thousand years. Dittmer concurs with other geographers in suggesting that further attention to the emotional dimensions of geopolitics may open new political possibilities.

The chapter by Richard Phillips explores geographical dimensions to the organizations and networks that have emerged in the UK in opposition to the war on terror and the invasion of Iraq in particular. Drawing on a wide range of interviews, he considers how different parts of the anti-war movement have situated themselves within traditions of anti-imperialism and examines the metaphorical and material geographies involved in their opposition to the invasion of Iraq and the war on terror. The chapter illuminates the geographies of protest and networking that enabled a range of organizations (including Stop the War, the Muslim Association of Britain and the Scottish Nationalist Party) to come together in shared opposition. The chapter suggests that a consideration of the complex geographies of empire and colonialism enabled otherwise diverse groups to find common ground for political action. Phillips suggests that while the anti-war movement may have failed in its own terms, new connections and understandings arising out of the geographies of resistance may create new resources for thinking beyond the war on terror.

The final chapter, by Alan Ingram, explores points of convergence between critiques of the war on terror in geography and in a range of artistic interventions and suggests that there is considerable scope for further engagement between the two. The chapter takes as its empirical focus an exhibition (entitled *Green Zone/Red Zone*) that took place at Gemak, a new cultural institution located in The Hague. Drawing on discussions with Gemak's curator and with contributing artists, Ingram suggests that the exhibition constituted a powerful counter-mapping to dominant imaginative geographies, and proposed connections and a sense of responsibility that cut across geopolitical dividing lines. Though he is sanguine about the ability of art to effect social change, Ingram suggests that art practices such as those in evidence at *Green Zone/Red Zone* represent an intriguing and important entry point for geographers concerned with the war on terror and landscapes of security.

As we have noted, this book, like all others, is a form of situated knowledge that, while it stretches more widely, reflects its production from within the

UK, the particular entanglement of that country in the war on terror and its privileged location in the production of imaginative geographies. It claims to be neither comprehensive nor systematic, but through suggestive illustrations aims to provide a richer sense of the multiple geographies of the war on terror and emerging landscapes of security, of their contradictions, and of the different ways in which people are responding to them. It is hoped that the book will become a resource for researching the landscapes of geopolitics and security, and that by counter-posing reflexive geographical imaginations and critiques to the dominant geopolitics of the war on terror it may inspire or provoke further geographies from other locations.

Acknowledgements

We thank Jason Dittmer, Stuart Elden, Alex Jeffrey and Merje Kuus for their helpful comments on an earlier draft of this introduction.

PART 1
Constructing the War on Terror

Chapter 2
Blair, Neo-Conservatism and the War on Territorial Integrity[1]

Stuart Elden

Introduction

Former British Prime Minister Tony Blair has been one of the most forceful advocates for a greater role for the 'international community' in the domestic affairs of sovereign states. In this regard it is perhaps entirely appropriate that he should be named as the special envoy for one configuration of the 'international community' – the quartet of the UN, US, Russia and the EU – for the Middle East, though the aim of creating peace in the region can only be shockingly inappropriate, given his role in the Iraq war. The disjuncture is one of the key themes explored in this paper, looking at the legal position for British involvement in that war.[2]

Blair's calls for the role of the international community were first made around the time of the Kosovo conflict, and were later used to justify actions taken in East Timor and Sierra Leone. Blair stated in 2001 that, 'if Rwanda happened again today ... we would have a moral duty to act there also'.[3] There is an explicit relation to the positions earlier advocated by British Foreign Secretary Robin Cook calling for an 'ethical dimension' to foreign policy (1997).[4] Neither individual was alone in this, as they were trading on fairly long standing arguments about humanitarian intervention. Yet at the same time rather different strands of opinion in the US were making not dissimilar claims about 'contingent' sovereignty, particularly in the Project for the New American Century's (PNAC) report of 2000, *Rebuilding America's Defenses*.[5] Since the terrorist attacks of 11 September

---

1 A slightly different and longer version of this essay was published as 'Blair, Neo-conservatism and the War on Territorial Integrity', *International Politics* 44 (2007) 37–57. Reprinted with permission from Palgrave Macmillan.

2 For a reading of the US legal position, see J. Prados, *Hoodwinked: The Documents that Reveal How Bush Sold Us a War* (New York: The New Press 2004).

3 T. Blair, 'Speech to the Labour Party Conference' (2 October 2001) at <http://politics.guardian.co.uk/labour2001/story/0,1414,562006,00.html>.

4 R. Cook, 'Speech on the Government's Ethical Foreign Policy' (12 May 1997) at <http://www.guardian.co.uk/ethical/0,2759,181072,00.html>.

5 Project for the New American Century, *Rebuilding America's Defenses: Strategy, Forces and Resources for a New Century* (2000) at <http://www.newamericancentury.org/RebuildingAmericasDefenses.pdf>.

2001, these two groupings – which might be supposed to be rather distinct – have come together in an alliance that has made an explicit challenge to long-standing norms of international law and international politics. These are the notion of equal sovereignty of states and the principle of non-intervention in domestic affairs. Both of these norms relate to the legal concept of territorial integrity, both in practice, and particularly in the UN Charter, which rests on three mutually enforcing ideals – the sovereign equality of all states; internal competence; and preservation of existing boundaries.

Although these ideals are regularly violated, the semblance of order in the international system requires them to operate as founding principles, or as necessary myths. That they have been challenged since 11 September 2001 is not especially remarkable, but what should give us cause to pause is the way in which they have been explicitly argued against. This chapter contends that these arguments can be profitably analysed and assessed through the lens of the issue of territorial integrity, and this is illustrated through a reading of the advice given to the British government over the legality of the war on Iraq. In doing so, this chapter seeks to analyse the convergence between the ethical aspects of foreign policy and the call for a notion of 'international community' by Blair on the one hand, and the project of the neo-conservatives in the US on the other. In so doing it illustrates the convergence and divergence between two dominant strands of thinking in world politics, and begins to interrogate the legal status of intervention. The chapter therefore offers a way into interrogating the international legal aspects of contemporary geopolitics, as well as offering a contribution to a geographical reading of international law.

Territorial Integrity and Humanitarian Intervention

Territorial integrity is a complicated notion in international law, as it has two distinct yet usually compatible meanings. One is that states should not promote secessionist movements in other states, nor try to seize land from them. This is the idea of territorial preservation, the continuation of existing boundaries and the cementing of the territorial status quo. On decolonization in South America or Africa, for example, states inherited the boundaries of colonial divisions, a notion legally known as *uti possidetis*.[6] The International Court of Justice has claimed that this is not a particularity of those cases, but a 'principle of general scope, logically connected with the phenomenon of the obtaining of independence, wherever it occurs'.[7] The second meaning is that within this territory, within its boundaries,

---

6   S. Lalonde, *Determining Boundaries in a Conflicted World* (Montreal and Kingston: McGill-Queen's University Press 2002).

7   International Court of Justice, 'Case concerning the frontier dispute Burkina Faso/ Republic of Mali' (22 December 1986) at <http://www.icj-cij.org/icjwww/icases/iHVM/ ihvm_summaries/ihvm_isummary_19861222.htm>.

the state is sovereign. This trades on the idea of equal sovereignty and accepts what the EU calls internal competence. Of course, no state is absolutely sovereign, both in terms of the powers held by other power groupings within its boundaries, and whole rafts of international law limit a state's competence in myriad ways. Yet for actions which do not have an effect beyond its borders, a state has been held to be sovereign, the notion of territorial sovereignty.[8]

In Walzer's work on just war theory, the support for 'the rights of political communities ... territorial integrity and political sovereignty' is important. But for Walzer, while these 'belong to states ... they derive ultimately from the rights of individuals', namely life and liberty, and they can therefore be defended on the same basis. It is on this basis that 'every violation of the territorial integrity or political sovereignty of an independent state is called aggression'.[9]

What this means – and this is important for some of the justifications used for the violation of territorial integrity – is that if the state fails in its contract with the people, interventions can sometimes be ethically justified, even though 'the practice of intervening often threatens the territorial integrity and political independence of invaded states'. The grounds for this are threefold: where the issue is secession or 'national liberation' for a particular community within a set of boundaries; counter-intervention to protect boundaries that have already been crossed; and where there is terrible 'violation of human rights', such as 'cases of enslavement or massacre'. These would be just wars 'that are not fought in self-defence or against aggression in the strict sense'.[10]

Secession is necessarily a challenge to territorial integrity. Yet without territorial integrity, Buchanan claims, states are not only not able to survive, but they are also not able to discharge their responsibilities to the 'most basic morally legitimate interests of the individuals and groups that states are empowered to serve, their interest in the preservation of their rights, the security of their persons, and the stability of their expectations'. States therefore do not merely have a 'morally legitimate interest in maintaining the principle of territorial integrity', but an '*obligatory* interest'.[11]

However, Buchanan similarly wants to challenge the absolutist interpretation of territorial integrity, where it applied to all states, and proposes a more circumscribed version, which applied to legitimate states only. This is what he calls 'the morally

---

8   For fuller analyses see S. Akweenda, 'Territorial Integrity: A Brief Analysis of a Complex Concept', *Revue Africaine de Droit International et Comparé* 1 (1997) 500–506; S. Elden, 'Territorial Integrity and the War on Terror', *Environment and Planning A* 37/12 (2005) 2083–104; and M. Zacher, 'The Territorial Integrity Norm: International Boundaries and the Use of Force', *International Organization* 55/2 (2001) 215–30.

9   M. Walzer, *Just and Unjust Wars* (New York: Basic Books 1992) pp. 52–4 and pp. 61–2.

10   Ibid., p. 86 and p. 90.

11   A. Buchanan, 'Theories of Secession', *Philosophy and Public Affairs* 26 (1997), p. 47.

progressive interpretation of the principle of territorial integrity'.[12] States are not legitimate if they 'threaten the lives of significant portions of their populations by a policy of ethnic or religious persecution' or if they deprive 'a substantial proportion of the population of basic economic and political rights'. The first case is exemplified by the infringement of 'Iraq's territorial integrity in order to establish a 'safe zone' in the North for Kurds' and the second is illustrated by the conditions under *apartheid* South Africa.[13] This is the basis for the argument for humanitarian intervention, where the failure of a state to discharge its responsibilities to its populations can legitimate international intervention. Indeed, Walzer goes further, suggesting that wars can be just if they are to support representative secessionist movements, to balance out another state's intervention, or for humanitarian reasons. For Walzer, moreover, 'we permit or, after the fact, we praise or don't condemn these violations of the formal rules of sovereignty, because they uphold the values of individual life and communal liberty of which sovereignty itself is merely an expression'.[14]

There is, of course, an unexamined 'we' here that calls for, legitimates and undertakes forms of international intervention. In some of the recent calls for the 'responsibility to protect' (only partly adopted), the answer is a reformed UN Security Council.[15] For Blair the solution is the 'international community', which may or may not coincide with the will of the Security Council. In the latter case, the charge is obviously that states are acting in their own interests, but for Walzer this is not necessarily a problem. In fact he declares that 'mixed motives are a practical advantage', because a combination of acting 'in their own interests as well as in the interests of humanity' means action is taken.[16] This parallels Blair's assertion of a notion of 'enlightened self-interest', where he claimed that, 'self-interest and our mutual interests are today inextricably woven together'.[17] Indeed, Blair even claimed that such rhetorical tropes could explain more than this: 'politics is different in America. This is a Republican administration with a certain view, so they will couch what they do in terms of US national interest, not international community. But the doctrine of international community is just enlightened national self-interest, so whatever the different rhetorical perspectives you come to the same point'.[18]

---

12   Ibid., p. 50.
13   Ibid., p. 50.
14   Walzer, *Just and Unjust Wars*, p. 108.
15   S. Elden, 'Contingent Sovereignty, Territorial Integrity and the Sanctity of Borders', *SAIS Review of International Affairs* 26/1 (2006) 11–24.
16   M. Walzer, 'The Argument About Humanitarian Intervention', *Dissent* 49/1 (2002) 29–37.
17   T. Blair (2002b), 'The Power of World Community', in M. Leonard (ed.), *Re-ordering the World: The Long-Term Implications of September 11th* (London: The Foreign Policy Centre 2002) 119–24, p. 120.
18   D. Goodhart, 'Tony's World', *Prospect* 77 (August 2002) at <http://www.prospect-magazine.co.uk/printarticle.php?id=5347&category=151&issue=484&author=&AuthKey

It is worth a little more examination of the background. In its 1997 election manifesto Labour declared that it wanted 'Britain to be respected in the world for the integrity with which it conducts its foreign relations', with the 'promotion of human rights' and the 'creation of a permanent international criminal court' to be priorities.[19] Rather than the later media shorthand of 'ethical foreign policy', Cook's early speeches noted an 'ethical content', suggesting that foreign policy should have 'an ethical dimension'.[20] Although Number 10 initially showed some caution, Blair later did not want the ethical dimension to be confined to the arms trade, but to include environmental issues, crime and – tellingly – 'the right to secure frontiers'.[21] One of the first tests of this came in the Balkans, in Kosovo, with the former Supreme Allied Commander Europe, Wesley Clark, noting the importance of Blair's election in changing strategic priorities in that region, and with the pressure for ground troops in the Kosovo war particularly coming from him.[22]

For Stevens, this marked a profound shift, in that NATO's first war 'was not being fought over territory but to uphold a set of values' and because it challenged the 'international system that had prevailed since the founding of the United Nations', namely the idea of non-interference in the actions of states within their own territory.[23] However, the relation is more complicated than this. While Blair too stressed that this was 'a just war, based not on any territorial ambitions but on values',[24] his reference to frontiers is essential: this war was inherently about territorial integrity. On the one hand Blair sought to challenge the notion of internal competence, but at the same time insisted on the territorial integrity (that is, preservation, of 'secure frontiers') of Yugoslavia in several international forums, including the G8, at Rambouillet and in the UN Security Council. But the challenge to territorial sovereignty was clear, and indeed, the Yugoslav government protested to the UN that its territorial integrity was being violated. Blair stated the inherent tension, and the territorial aspect, of non-intervention in a famous speech in Chicago. Here he suggested that we should not 'jettison too readily' this principle, as 'one state should not feel it has the right to change the political system of another or foment subversion or seize pieces of territory to which it feels it should have some claim'. But this did not mean that it should not be

=6a507cc8bf8575be654eac84c8d7b79d>.

19   Labour Party, *Labour Party Manifesto* (London: Labour Party 1997).

20   Cook (note 4).

21   J. Kampfner, *Robin Cook* (London: Phoenix) p. 216.

22   W. Clark, *Waging Modern War: Bosnia, Kosovo and the Future of Combat* (New York: Public Affairs 2002).

23   P. Stevens, *Tony Blair: The Making of a World Leader* (New York: Viking 2004) p. 162; see p. 170.

24   T. Blair, 'Speech to the Chicago Economic Club' (22 April 1999) at <http://www.number–10.gov.uk/output/Page1297.asp>.

'qualified in important respects'.[25] These qualifications were genocide; oppression leading to refugees; and minority rule. Despite these, the broad scope was evident. For Lord Robertson, formerly Secretary of State for Defence under Blair and then NATO Secretary-General, the Chicago speech did for foreign politics what reform of Clause IV did domestically: 'it was one of those occasions where Tony Blair set out a bold line which changed the whole geography'.[26]

For Blair, the 'doctrine of international community' concerned 'a community based on the equal worth of all, on the foundation of mutual rights and mutual responsibilities'.[27] After 11 September 2001, Blair would claim that even before he was 'already reaching for a different philosophy in international relations from a traditional one that has held sway since the treaty of Westphalia in 1648; namely that a country's internal affairs are for it and you don't interfere unless it threatens you, or breaches a treaty, or triggers an obligation of alliance'.[28] This needed to be done by Europe and America, who should jointly push for 'a greater role of leadership for the UN on the responsibility of states to protect not injure their own citizens'.[29]

Despite continuity, it is important to note a crucial difference. In 2004 Blair followed his call for UN reform by suggesting that, 'none of this will work, however, unless America too reaches out. Multilateralism that works should be its aim. I have no sympathy for unilateralism for its own sake'.[30] But in 1999 he had noted something rather different. Instead of the US reaching out to the world, the world needed the US, as 'those nations which have the power, have the responsibility. We need you engaged. We need the dialogue with you'. This was followed by a plea for avoiding 'the doctrine of isolationism' as the world could not afford it.

The Neo-Conservative Challenge

Changing administrations in the US have shown both parallels and significant tensions. As Clinton declared in his first inaugural address, 'when our vital interests are challenged, or the will and conscience of the international community is defied, we will act; with peaceful diplomacy whenever possible, with force when

---

25   Ibid.

26   Cited in A. Seldon, *Blair* (London: Free Press 2005) p. 407.

27   T. Blair, 'Speech to the Global Ethics Foundation, Tübingen University, Germany' (30 June 2000) at <http://www.number-10.gov.uk/output/Page1529.asp>.

28   T. Blair, 'Speech in Sedgefield' (5 March 2004) at <http://politics.guardian.co.uk/iraq/story/0,12956,1162991,00.html>.

29   T. Blair, 'Mansion House Speech on Foreign Policy' (16 November 2004) <http://politics.guardian.co.uk/foreignaffairs/story/0,11538,1352442,00.html>.

30   Ibid.

necessary.[31] The significance of the phrase 'vital interests' is important, as is the reference to the notion of 'international community'. Tellingly, Clinton suggested to the UN that the US would act 'multilaterally when possible, but unilaterally when necessary'. As Derrida notes, it is significant that this claim was in relation to article 51 of the UN founding Charter, that is 'the article of exception'.[32]

This article makes clear that nothing in the Charter limits the right of 'individual or collective self-defence' (the exception), but this itself is limited to it being in response to 'an armed attack', and until the Security Council 'has taken measures necessary to maintain international peace and security'. But even under the Clinton administration (particularly in its second term) this willingness to act was not confined to armed attack but was understood more generally. As Derrida recounts, Secretary of Defense William S. Cohen was willing to 'intervene militarily in a unilateral way (and thus without the prior accord of the United Nations or the Security Council) each time its vital interests were at stake; and by vital interests he meant 'ensuring uninhibited access to key markets, energy supplies, and strategic resources', along with anything that might be considered a 'vital interest by a "domestic jurisdiction"'.[33]

What we find is thus not a dramatic change in the transition from Clinton to Bush, but at most a shift of emphasis in relation to the Clinton administration's version of internationalism. Cohen was a Republican member of Senate before his appointment for the second Clinton term, and had made clear his support for a bipartisan national security policy. The Clinton *National Security Strategy of Engagement and Enlargement*, for example, shows evidence of much of what followed.[34] But if in this earlier period the emphasis had been on unilateralism as a last resort, with a multilateral strategy preferred,[35] the shift was to one where multilateral strategies were pursued only if felt necessary. Clinton's 'new internationalist foreign policy' was designed to make the US 'the "indispensable nation"' in such issues as peace in Europe, Northern Ireland, and the Middle East and to 're-establish collective security for a new age of globalisation and interdependence'.[36]

---

31   W. Clinton, 'First Inaugural Address' (20 January 1993) at <http://www.unclefed. com/EduStuff/HistDocs/clinton1.html>.

32   J. Derrida, *Rogues: Two Essays on Reason*, translated by Pascale-Anne Brault and Michael Nass (Stanford: Stanford University Press 2004) p. 103.

33   Derrida (ibid.) pp. 103–4, citing N. Chomsky, *Rogue States: The Rule of Force in World Affairs* (Cambridge MA: South End Press 2000) p. 4.

34   The White House, *A National Security Strategy of Engagement and Enlargement* (Washington DC: White House 1995) at <http://www.dtic.mil/doctrine/jel/research_pubs/ nss.pdf>; 'J. Der Derian, 'Decoding the National Security Strategy of the United States', *Boundary* 2/30 (2005) 19–27.

35   P. Riddell, *Hug Them Close: Blair, Clinton, Bush and the 'Special Relationship'* (London: Politico's 2003) p. 59.

36   S. Blumenthal, *The Clinton Wars* (New York: Plume Books 2003) p. 789.

The reaction to this internationalism within neo-conservative thinking was pronounced. One of the issues was the supposed surrender of US interests to the UN, along with the critique of Madeleine Albright's notion of 'assertive multilateralism', by which is usually meant a majority in the Security Council, minus any 'great'-power veto. In distinction, as Mann puts it, the new Bush regime had 'a unilateralist and militarist vision of how to overcome world disorder'.[37] Charles Krauthammer calls this a form of realism, a 'new unilateralism', an explicit recognition of unipolarity. But he cautions this is not isolationism, because 'the new unilateralism defines American interests far beyond narrow self-defence. In particular it identifies two other major interests, both global: extending the peace by advancing democracy and preserving the peace by acting as balancer of last resort'.[38]

Alongside this willingness to go against the UN or indeed world opinion more generally was a swift reaction against multilateral treaties.[39] As Lind notes, 'in his first year, Bush cancelled more international treaties than any president in American history'.[40] Not only did Bush refuse to ratify Kyoto, but the US walked out of the Durban conference against racism for criticisms of Israel, and refused to be a party to the International Criminal Court (ICC), or to conventions, protocols and treaties on the rights of the child, landmines, and biological weapons, and unilaterally withdrew from the anti-ballistic missile treaty.[41] John Bolton's appointment as ambassador to the UN, and the 750 changes he proposed to the 2005 World Summit outcome document are further indications of this resistance to multilateralism.[42]

This was largely the basis on which Bush had run for president, with an intention to take a 'a narrow view of America's national interest'. Bush and Condoleezza Rice had indicated that they were 'not interested in do-gooding in far-flung lands', and were therefore 'scornful of the multilateralism at the heart of Blair's outlook'.[43] 11 September 2001 was the catalyst for a rethinking. For Blair, these

---

37   M. Mann, *Incoherent Empire* (London: Verso 2003) p. 2; Derrida (note 33) pp. 95–6; p. 98.

38   C. Krauthammer, 'The Unipolar Era', in A. Bacevich (ed.), *The Imperial Tense: Prospects and Problems of American Empire* (Chicago: Ivan R. Dee 2003) 47–65, p. 60.

39   C. Krauthammer, *Democratic Realism: An American Foreign Policy for a Unipolar World* (Washington DC: The AEI Press 2004) pp. 5–6.

40   M. Lind, *Made in Texas: George W. Bush and the Southern Takeover of American Politics* (New York: Basic Books 2004) p. 134.

41   See N. Smith, *The Endgame of Globalization* (London: Routledge 2005) p. 193; Z. Brzezinski, *The Choice: Global Domination or Global Leadership* (New York: Basic Books 2004) p. 230 (note 6); S. Halper and J. Clarke, *America Alone: The Neo-Conservatives and the Global Order* (Cambridge: Cambridge University Press 2004) pp. 122–9.

42   Elden, 'Contingent Sovereignty' (note 15).

43   Stevens, *Tony Blair* (note 23) p. 188; see C. Rice, 'Campaign 2000: Promoting the National Interest', *Foreign Affairs* 79/1 (2000) at <http://www.foreignaffairs.org/ 20000101faessay5/condoleezza-rice/campaign-2000-promoting-the-national-interest.

events meant that his earlier themes of 'international community' could be seen in sharper focus.[44] And in the immediate aftermath around the issue of Al Qaeda and the Taliban in Afghanistan there would indeed be international support. 'Nous sommes tous américains' as *Le Monde* famously declared, and NATO invoked Article 5, declaring these were attacks on all members. Superficially there was a shift in the Bush rhetoric.

> Our nation's cause has always been larger than our nation's defense. We fight, as we always fight, for a just peace – a peace that favors human liberty. We will defend the peace against threats from terrorists and tyrants. We will preserve the peace by building good relations among the great powers. And we will extend the peace by encouraging free and open societies on every continent.[45]

Indeed this and the mobilization of the figures of 'good' and 'evil' show how the religious aspects of Bush's conservatism, which in the 2000 election had largely been confined to domestic issues, now took on a much wider international 'moral' agenda.[46] While this seems to mirror Blair's intentions, it masks a much harder line of the unipolar moment in US foreign policy, which refused to be bound by the collective constraints of NATO[47] or the UN. Elements within the Bush administration used 11 September 2001 as an opportunity for implementing ideas dating back to Cheney and Wolfowitz's 'Defense Planning Guidance' of 1992, to PNAC's *Rebuilding America's Defenses* (2000), and a report written for the Harvard Visions of Governance for the Twenty-First Century project by Ashton B. Carter, John M. Deutch and Philip D. Zelikow, which had declared that

> International norms should adapt so that such states are obliged to reassure those who are worried and to take reasonable measures to prove they are not secretly developing weapons of mass destruction. Failure to supply such proof, or prosecute the criminals living in their borders, should entitle worried nations to take all necessary actions for their self-defense.[48]

---

html>; Halper and Clarke, *America Alone* (note 41) pp. 134–5; J. Naughtie, *The Accidental American: Tony Blair and the Presidency* (London: Macmillan 2004) p. 73.

44 Stephens, *Tony Blair* (note 23) p. 201.

45 G. Bush, 'Remarks by the President at 2002 Graduation Exercise of the United States Military Academy' (1 June 2002) at <http://www.whitehouse.gov/news/releases/2002/06/20020601-3.html>.

46 Lind (note 40) and T. Frank, *What's the Matter with Kansas? How Conservatives won the Heart of America* (New York: Owl Books 2005).

47 W. Clarke, *Winning Modern Wars: Iraq, Terrorism and the American Empire* (New York: Public Affairs Press 2003) pp. 126–8.

48 A. Carter, J. Deutch and P. Zelikow, *Catastrophic Terrorism: Elements of a National Policy* (Cambridge MA: Visions of Governance for the Twenty-First Century 1998) at <http://www.ksg.harvard.edu/visions/publication/terrorism.htm>.

The policies enshrined in the National Security Strategy of 2002 and the subsequent strategies outline ways in which territorial sovereignty is open to challenge. But at the same time they continually reinforce the importance of states being fully in control of their territory, the importance of territorial stability and the dangers of failed states (see Chapter 3).[49]

The foreign policy projects of Blair and Bush thus explicitly demand changes to the notion of territorial integrity in its second sense of territorial sovereignty, while reinforcing the first sense of territorial integrity as preservation of existing boundaries. Sovereignty is seen as contingent within the fixed boundaries of nation-states. The projects are not entirely compatible, however, for there is one key distinction, which can only be simply rendered as multilateralism versus unilateralism, and might better be understood as the question of internationalism for Blair, versus the uniqueness or exceptionalism of the US. Both tensions – between Blair and the US, and between territorial sovereignty and territorial preservation – were also at the heart of the debate over Iraq.

Sovereignty, Territory and the Challenge to the UN

The justification for an attack on Iraq took many forms, and although the question of weapons of mass destruction took the headlines, other reasons, including Saddam's threat to his neighbours and putative links to terrorism, and his undeniable human rights abuses were also part of a confused – and intentionally confusing – rationale. These reasons were, according to Paul Wolfowitz, all at stake, even though the weapons issue was privileged 'for reasons that have a lot to do with the US government bureaucracy'.[50] There is not the space here for a full discussion of the reasons given, but a few points are worth making.

Just as in Afghanistan, there was an attempt to tie the global security issues to the basis of the internal actions of the regimes they aimed to depose – counter-terrorism as humanitarian intervention. In addition, by the beginning of the war, Saddam was clearly not in control of all of Iraq's territory, nor could he guarantee the needs of its population. Iraq was, in these terms, a 'failed state'.[51] What these necessarily conflate is an external threat and internal actions. While a right to self-preservation in the face of a forthcoming attack is legitimated under international law, and can – in tightly circumscribed ways – allow the violation of another

---

49   The White House, *The National Security Strategy of the United States* (Washington DC: The White House 2002) at <http://www.whitehouse.gov/nsc/nss.html>.

50   P. Wolfowitz, 'Interview with Sam Tannenhaus', *Vanity Fair* (9 May 2003) at <http://www.defenselink.mil/transcripts/2003/tr20030509-depsecdef0223.html>. For a critical reading see J. George, 'Leo Strauss, Neo-Conservatism and US Foreign Policy: Esoteric Nihilism and the Bush Doctrine', *International Politics* 42/2 (2005) 174–202.

51   L. Anderson and G. Stansfield, *The Future of Iraq: Dictatorship, Democracy or Division?* (London: Palgrave Macmillan 2004) pp. 83–4.

state's territorial integrity, this is not the case for internal actions. What we find is an attempt to use the internal actions of the regime as a part justification for intervention, while at the same time denying this is the case.[52]

The claims of the US go further than this. In fact, they claim a right to pre-empt danger, that is to take action before threats materialize. Legal advice to the US Congress laid this out, suggesting that Iraq could not be presented as an 'imminent threat' that would justify pre-emption, *except* on two bases: possession of WMD and links to terrorist groups that might use them against the US. It suggested that this necessarily related to the National Security Strategy question of whether pre-emption 'ought to be recast in light of the realities of WMD, rogue states, and terrorism'.[53] British Attorney General Lord Goldsmith noted that, 'this is not a doctrine which, in my opinion, exists or is recognised in international law'.[54] But as well as being illegal, it is also potentially counter-productive, in that states that may find themselves potential US targets are likely to want the same capacity to retaliate in advance as the US currently has. It is notable that Bolton removed a UN call for 'nuclear weapons States to reaffirm their commitment to Negative Security Assurances' from the World Summit outcome document in 2005. This means that the US reserves the right to first-strike a non-nuclear power, outside even of the initial restrictions to this policy. For Bolton the document was flawed as it 'emphasizes disarmament, when the true threat to international security stems from proliferation'.[55] As Maogoto phrases it, 'what Bush fails to realise is that his actions will encourage other states to acquire the very weapons that he purports to abhor'.[56] But it is clear from the security strategies put forward by the US administration that they have anticipated this, and that one of their aims is to prevent other states gaining that capacity to defend themselves from a US pre-emptive attack, lest the US lose its power of deterrence.

On 23 July 2002, Blair met with the Defense Secretary, the Foreign Secretary, the Attorney-General, military and intelligence officials and advisors. The leaked

---

52   N. Maogoto, 'New Frontiers, Old Problems: The War on Terror and The Notion of Anticipating the Enemy', *Netherlands International Law Review* LI (2004) 1–39. See also C. Greenwood, 'International Law and the "War on Terrorism"', *International Affairs* 78/2 (2002) 301–17.

53   D. Ackermann, 'International Law and the Preemptive Use of Force Against Iraq' (Washington DC: CRS Report for Congress 2003) at <http://www.usembassy.it/pdf/other/RS21314.pdf> p. 6; The White House, *National Security Strategy* 2002 (note 49) p. 15.

54   P. Goldsmith, 'Memo to Prime Minister on Iraq: Resolution 1441' (7 March 2003) at <http://image.guardian.co.uk/sys-files/Guardian/documents/2005/04/28/legal.pdf> p. 3 and C. Greenwood, 'Memorandum: The Legality of Using Force Against Iraq', Select Committee on Foreign Affairs Minutes of Evidence at <http://www.publications.parliament.uk/pa/cm200203/cmselect/cmfaff/196/2102406.htm> p. 24.

55   J. Bolton, 'Letter to the United Nations and Enclosure on United States Amendments' (1 September 2005) p. 2.

56   Maogoto (note 52) p. 39.

minute of that meeting is revealing.[57] It reports on the intelligence advice, which notes that Washington now saw military action as 'inevitable', with no patience for the UN route, and with justification on the basis of 'the conjunction of terrorism and WMD'. It was reported that, 'the intelligence and facts were being fixed around the policy'. The opinion of Jack Straw, Foreign Secretary, who was shortly to discuss this with Colin Powell was that 'the case was thin', as Saddam was not 'threatening his neighbours, and his WMD capability was less than that of Libya, North Korea or Iran'. Straw urged planning for 'an ultimatum to Saddam to allow back in the UN weapons inspectors', that 'would also help with the legal justification for the use of force'. In 2002 then a case was not felt to exist, which at the very least complicates the continuation or revival argument later used. Indeed, the opinion of the Attorney-General shows the legal situation was already being worked out, and that 'regime change was not a legal base for military action'. He reported that

> There were three possible legal bases: self-defence, humanitarian intervention, or UNSC authorisation. The first and second could not be the base in this case. Relying on UNSCR 1205 of three years ago would be difficult. The situation might of course change.[58]

The meeting closed with the clear direction that 'we should work on the assumption that the UK would take part in any military action', but that they 'must not ignore the legal issues: the Attorney-General would consider legal advice with FCO/MOD legal advisers'.[59]

The background to this meeting is important. Blair had told Bush at his Crawford Ranch in April 2002 'that the UK would support military action to bring about regime change',[60] despite the fact that as all advice states, and the Attorney General reinforces later, this is not permissible. What is telling about this part of his legal advice is that it lays out the limits of actions in war: not merely legal, but ethical in the way it relates to criteria for just war. The objective must be enforcing the ceasefire of resolution 687 (1990); the scope must be limited to that; and the means used must be proportionate. This leaves open the possibility of removing 'Saddam Hussein from power if it can be demonstrated that such action is a necessary and proportionate measure to secure the disarmament of Iraq'. This is why regime change in itself 'cannot be the objective of military action', and

---

57   S. Rycroft, 'Iraq: Prime Minister's Meeting, 23 July 2002, Memo to David Manning, S 195/02', *The Sunday Times* (1 May 2005) at <http://www.timesonline.co.uk/article/0,,2087-1593607,00.html>.

58   Ibid.; Goldsmith, 'Resolution 1441' (note 54) p. 2.

59   Rycroft (note 57).

60   Memo of 19 July 2002, reported in R. Norton-Taylor and P. Wintour, 'Papers Reveal Commitment to War', *The Guardian* (2 May 2005) at <http://politics.guardian.co.uk/election/story/0,15803,1474755,00.html>.

why 'this should be borne in mind in considering the list of military targets and in making public statements about any campaign'.[61]

In March 2002 a series of memos were produced for this meeting between Blair and Bush. A number of these can be found in a valuable collection of documents.[62] One is especially important, since it accepts that 'a legal justification for invasion would be needed', and that 'subject to Law Officers' advice, none currently exists'.[63] It then sets out the 'Current Objectives of UK Policy'. The argument is made that there are two options: 'A toughening of the existing containment policy, facilitated by 11 September; or regime change by military means: a new departure which would require the construction of a coalition and a legal justification'.[64] The outcomes part of this options paper outlines two possible 'end states' – 'a Sunni strongman or a representative government'.[65] The first, it is suggested, 'would be likely to maintain Iraqi territorial integrity', and would allow a quick troop withdrawal but could create the same dangers as Saddam, even including WMD.[66] The second is inherently more risky, creating a

> representative broadly democratic government. This would be Sunni-led but within a federal structure ... Such a regime would be less likely to develop WMD and threaten its neighbours. However, to survive it would require the US and others to commit to nation building for many years. This would entail a substantial international security force and help with reconstruction.[67]

The memo's conclusion is that

> a full opinion should be sought from the Law Officers if the above options are developed further. But in summary: CONTAINMENT generally involves the implementation of existing UNSCRs and has a firm legal foundation. Of itself, REGIME CHANGE has no basis in international law.[68]

It thus tries to link the latter to the former, suggesting that it is one of the mechanisms of best enforcing previous resolutions. As then Foreign Secretary Jack Straw wrote to Blair on 25th March 2002:

---

61   Goldsmith, 'Resolution 1441' (note 54) p. 36.

62   M. Danner, *The Secret Way to War: The Downing Street Memo and the Iraq War's Buried History* (New York: New York Review Press 2006); T. Simpson (ed.), *The Dodgiest Dossier* (Nottingham: Spokesman 2005).

63   Danner (note 62) p. 96.

64   Ibid., p. 99.

65   Ibid., p. 95.

66   Ibid., p. 103.

67   Ibid., p. 104.

68   Ibid., p. 112.

Legally there are two potential elephant traps:

i. regime change per se is no justification for military action; it could form part of the method of any strategy, but not a goal. Of course, we may want credibly to assert that regime change is an essential part of the strategy by which we have to achieve our ends – that of the elimination of Iraq's WMD capacity; but the latter has to be the goal;

ii. on whether any military action would require a fresh UNSC mandate (Desert Fox did not). The US are likely to oppose any idea of a fresh mandate. On the other side, the weight of legal advice here is that a fresh mandate may well be required.[69]

The UK therefore tried to open up the possibility of the revival argument:

Currently, offensive military action against Iraq can only be justified if Iraq is held to be in breach of the Gulf War ceasefire resolution, 687. 687 imposed obligations on Iraq with regard to the elimination of WMD and monitoring these obligations. But 687 never terminated the authority to use force mandated in UNSCR 678 (1990). Thus a violation of 687 can revive the authorisation to use force in 678.[70]

Further it notes that as the ceasefire resolution (687) was proclaimed by the Security Council, it is therefore for the SC to decide if it has been breached, and therefore to revive 678. UNSCR 1205 (1998) is seen as a precedent. This was when Iraq had expelled weapons inspectors, and therefore was deemed to have breached 687. 'In our view, this revived the authority for the use of force under 678 and underpinned Desert Fox'.[71]

Anticipating the issues that would come to a head in the legal debate in the UK in March 2003, the memo recognizes that 'in contrast to general legal opinion, the US asserts the right of individual Member States to determine whether Iraq has breached 687, regardless of whether the Council has reached this assessment'.[72] Since no other state recognized this view, the UK government suggests that Desert Fox might be the best model, but it recognizes that 'our interpretation of resolutions 1205 was controversial anyway; many of our partners did not think the legal basis was sufficient as the authority to use force was not explicit. Reliance on it now would be unlikely to receive any support'.[73]

---

69   Ibid., p. 148.
70   Ibid., p. 113.
71   Ibid., p. 113.
72   Ibid., p. 113, p. 119.
73   Ibid., p. 102.

Stevens notes that one Downing Street aide described Bush in Afghanistan as 'unilateralist in principle but multilateralist in practice'.[74] Tellingly, for Iraq the reverse was true, not for Bush – who abandoned the principle for US self-interest pure and simple – but for *Blair*. As Stevens suggests, 'Blair was by instinct a multilateralist, but his commitment to international institutions was conditional – the United Nations, he would often say, could not be an excuse for inaction'.[75] Blair was instrumental, along with Powell, in suggesting the UN route be followed at all, with Cheney a key advocate of the exceptionalist route. In fact this mirrored a larger debate within the US administration on policy, with the signatories of the PNAC statement of principles of 1997 advocating an exceptionalist position against the more cautious multilateral approach. Toward the end of the Clinton administration Wolfowitz and Stephen J. Solarz had called on the US government 'to commit ground forces to protect a sanctuary in southern Iraq where the opposition could safely mobilise'.[76] This was also urged by Project for a New American Century in 2000.[77] The balance undoubtedly shifted after the 'new Pearl Harbor' that PNAC had anticipated, in favour of a more clearly orientated unipolar approach.

While Riddell is correct that Blair's moral attitude does not make him a neo-conservative because 'his doctrine of humanitarian intervention is rooted in liberal values, international treaties and institutions, not American hegemony', there is a similar logic at play.[78] This distinction is the unresolved problem, which lay at the heart of Blair's foreign policy.[79] In Stothard's account of the days around the start of the war, he claims that on 25th March 2003 Blair admitted 'that he has been 'uncomfortable, frankly' with the context and confines of international law and United Nations resolutions'.[80] While he attempted to provide international support for the US's unilateralism by tying it to the UN route, he ultimately failed to deliver the second Security Council resolution on Iraq which would have provided unambiguous international support. Faced with an impasse, he still went along with the US in invading and therefore violating a sovereign state's territorial integrity. For Blair, as he stressed many times, a decision *had to be made*. Tracing how that decision was made, working through the tangled arguments used to justify

74   Stevens, *Tony Blair* (note 23) p. 204.

75   Ibid., p. 208.

76   S. Solarz and P. Wolfowitz, 'How to Overthrow Saddam' (Letter to the editor), *Foreign Affairs* March/April (1999); Mann (note 37) p. 333.

77   Project for the New American Century, *Rebuilding America's Defenses* (note 5).

78   P. Riddell, *Hug them Close: Blair, Clinton, Bush and the 'Special Relationship'* (London: Politico's 2003) p. 289.

79   J. Kampfner, *Blair's Wars* (London: Free Press 2004) pp. 216–7.

80   P. Stothard, *Thirty Days: Tony Blair and the Test of History* (New York: Harper Collins 2003) p. 141.

the reconciliation of these positions is revealing both in terms of the tensions between Blair and the neo-conservatives and between territorial sovereignty and preservation.

Ten Days: Who Decides?

In order to investigate these two tensions further, the rest of this chapter concentrates on the two legal statements given by the British Attorney General. The first was a 36 paragraph 'Secret' briefing on 7th March 2003, the second the much shorter nice points given to Parliament on 17th March 2003.[81] Although it has been claimed that the new advice was written by Christopher Greenwood,[82] it must be noted that Greenwood's evidence to the Select Committee on Foreign Affairs[83] is considerably more careful an argument in terms of the revival case than that eventually presented by the British government.

The key issue at stake is precisely a decision of sovereignty: who decides? Who decides if Saddam is in material breach of resolution 1441? The British opinion was originally that it was the UN's job to decide if Saddam was in material breach; the US position was that individual states (that is, the US) could make that decision, as it was a 'matter of objective fact', but Goldsmith notes that he is 'not aware of any other state which supports this view', which confirms the view of the British government from 2002.[84] The Attorney General is therefore equivocal in his advice of 7th March 2003. Although he notes that 'I disagree, therefore, with those commentators and lawyers, who assert that nothing less than an explicit authorisation to use force in a Security Council resolution will be sufficient', he concluded that, 'if an assessment is needed of that situation, it would be for the Council to make it. A narrow textual reading of the resolution suggests that sort of assessment is not needed, because the Council has predetermined the issue. Public statements, on the other hand, say otherwise'.[85]

It is for this reason, that he suggested that, 'the safest legal course would be to secure the adoption of a further resolution to authorise the use of force'.[86] In other words, the decision is one that it is best to bind to the international process. He notes, however, that the 'arguments of the US Administration which I heard in

---

81   Goldsmith, 'Resolution 1441' (note 54) and P. Goldsmith, 'The Legal Basis for the Use of Force against Iraq' (17 March 2003) at <http://www.guardian.co.uk/Iraq/Story/0,2763,1471659,00.html>.

82   R. Cook, *The Point of Departure: Diaries from the Front Bench* (London: Pocket Books 2004) p. 344.

83   Greenwood, 'Legality of Using Force Against Iraq' (note 54).

84   Goldsmith 'Resolution 1441' (note 54) p. 9; see p. 1; H. Blix, *Disarming Iraq: The Search for Weapons of Mass Destruction* (London: Bloomsbury 2004) p. 274.

85   Goldsmith 'Resolution 1441' (note 54) p. 26.

86   Ibid., p. 27.

Washington' lead him to 'accept that a reasonable case can be made that resolution 1441 is capable in principle of reviving the authorisation in 678 without a further resolution'.[87] But, this is only 'sustainable if there are strong factual grounds for concluding that Iraq has failed to take the final opportunity … we would need to be able to demonstrate hard evidence of non-compliance and non-cooperation'. This 'matter of objective fact' (in the US sense) would be dependent on the views of UNMOVIC (the UN Monitoring, Verification and Inspection Commission) and the IAEA (the International Atomic Energy Agency), which are both UN bodies. As Goldsmith notes, at least in the first case, headed by Hans Blix, this is insufficient.[88] The question here is whether the US is bound to return to the UN Security Council, or if the decision can be made on the basis of 'strong factual grounds' alone. In the latter case, the issue is who is responsible for providing those grounds.

The point here is not just that the case was insufficient. On 7th March he notes the clear distinction between the US and British positions. 'The question is who makes the assessment of what constitutes a sufficiently serious breach. On the UK view of the revival argument (though not the US view) that can only be the Council, because only the Council can decide if a violation is sufficiently serious to revive the authorisation to use force'.[89] In the 17th of March advice to Parliament, Goldsmith declared that 'it is plain that Iraq has failed so to comply and therefore Iraq was at the time of Resolution 1441 and continues to be in material breach',[90] thus making the British government as arbitrator of this. In the House of Commons the next day Blair notes that Iraq's '8 December declaration is false. That in itself is a material breach'. And yet shortly later he notes that, 'had we meant what we said in Resolution 1441, the Security Council should have convened and condemned Iraq as in material breach'.[91] This is indicated, as a statement of fact, one that the UN should have merely endorsed, not decided. Indeed Goldsmith claims explicitly that 'Resolution 1441 would in terms have provided that a further decision of the Security Council to sanction force was required if that had been intended. Thus, all that Resolution 1441 requires is reporting to and discussion by the Security Council of Iraq's failures, but not an express further decision to authorise force',[92] which sidesteps the issue by pretending it does not exist.[93]

This hints at the fudge of 1441, where the agreement was reached through precisely *not* having an automatic trigger, yet not explicitly stating that a further resolution was needed. This was the balance between, effectively, France and the US. Both could justify the resolution as having met their goals and as a vindication

---

87   Ibid., p. 28.
88   Ibid., p. 29; see Blix (note 84) especially p. 172.
89   Goldsmith, 'Resolution 1441' (note 54) p. 17.
90   Goldsmith, 'Legal Basis' (note 81) point 7.
91   T. Blair, 'Prime Minister's Statement Opening Iraq Debate, House of Commons' (18 March 2003) at <http://www.number-10.gov.uk/output/Page3294.asp>.
92   Goldsmith, 'Legal Basis' (note 81) point 9.
93   Goldsmith, 'Resolution 1441' (note 54) p. 18.

of their policy. There were several examples of this, perhaps most evident in the substitution of 'serious consequences' for Saddam rather than empowering the UN to take 'all necessary means' to enforce the resolution.[94] Indeed, this is a good example of a favoured Blair negotiation strategy: that of constructive ambiguity. Blair had used this effectively in Northern Ireland and he and Clinton had also used this to find a diplomatic way through a complicated situation in Kosovo.[95] What this allows is for an intentionally vague or non-specific resolution of a problem that both or all sides can agree on – or even leaving it out entirely – while leaving the actual solution to a later date.

Goldsmith considers the various arguments quite carefully – noting that Foreign Secretary Jack Straw and UN Ambassador Jeremy Greenstock had informed him of the background of these negotiations[96] – and actually notes that not only did the French and Russians feel that Resolution 1441 was not an automatic trigger, but that while 'the US objective was to ensure that the resolution did not constrain the right of action which they believed they already had, our objective [i.e. the UK's] was to secure a sufficient authorisation from the Council in the absence of which we would have had no right to act'. Greenstock had unequivocally stated in a speech on the day 1441 was passed that 'there is no 'automaticity' in this Resolution. If there is further Iraqi breach of its disarmament obligations, the matter will return to the Council for discussion as required ... We would expect the Security Council then to meet its responsibilities'.[97]

The strongest argument Goldsmith then has is that the word 'consider' was introduced into Resolution 1441 'deliberately to indicate the need for a further discussion, but not a decision', at least partly because the US felt that a 'material breach' was a 'matter of objective fact and does not require a Security Council determination'. Greenstock's speech would, if not support, at least not exclude this interpretation since it stresses it will return for 'discussion as required', but only that there is an expectation, rather than a requirement, that it would make the decision. Tellingly, Goldsmith notes that, 'by contrast, the UK position taken on the advice of successive Law Officers, has been that it is for the Security Council to determine the existence of a material breach of the ceasefire'. However, the US was determined that 'the resolution should not concede the need for a second resolution. They are convinced that they succeeded'.[98]

---

94   See, for example, Naughtie, *The Accidental American* (note 43) pp. 145–6; Blix (note 84) p. 89 and B. Kendall, 'Show Down at the UN', in S. Beck and M. Downing (eds), *The Battle for Iraq: BBC News Correspondents on the War against Saddam and a* New World Agenda (London: BBC 2003) pp. 52–65.

95   Blumenthal (note 36) p. 643 and M. Albright, *Madam Secretary: A Memoir* (London: Pan 2003) p. 415–6.

96   Goldsmith, 'Resolution 1441' (note 54) p. 1.

97   Quoted in Danner (note 62) p. 18.

98   Goldsmith, 'Resolution 1441' (note 54) p. 22.

Essentially the Goldsmith opinion of the 7th of March says that the UK position is that only the Security Council can authorise force (international), and that this has been the 'advice of successive Law Officers' but by the 17th he claims that this is either a decision that can be made without the Security Council (unilateral) or does not need to be made by them as they merely need to 'consider'. As he notes, he was in discussion with the US before the 7th March[99] and apparently again before 17th March. As *The Observer* discovered, and which the British government confirmed, Goldsmith met with five of the Bush Administration's lawyers – Alberto Gonzales, William Taft IV, Jim Haynes, John Bellinger and John Ashcroft – on 11 February 2003.[100] As Goldsmith says in the 7th March advice, he was 'impressed by the strength and sincerity of the views of the US Administration which I heard in Washington on this point', but that he is necessarily 'reliant on their assertions for the view that the French (and others) knew and accepted that they were voting for a further discussion and no more', rather than hard evidence. He suggests that the legal status of the negotiations as evidence was 'very uncertain'.[101] But on 17th March Goldsmith puts this in a rather different way. Either Saddam was in breach as a matter of fact, and thereby force was justified under Resolution 1441; or the SC did not need to discuss this anyway. This effectively gave Bush, with or without Blair, the final decision.

Conclusion

The 17th of March advice served a number of purposes. It was the document seen by cabinet, given to Parliament, and also to the armed forces. As Admiral Sir Michael Boyce, then Chief of the Defence Staff argued, the need for a clear legal opinion arose when the British were starting to get troops in the area.[102] Boyce noted that his 'concern was always that the troops should feel absolutely confident that what they were doing was absolutely black-and-white legal ... I just wanted to make sure that if my soldiers went to jail and I did some other people go as well with me ... I had a perfectly unambiguous black-and-white statement saying it would be legal to operate if we had to'.[103] In providing that advice, Blair's government may have allowed the war to go ahead, but it exposes

---

99   Ibid., p. 1.

100   A. Barnett and M. Barrett, 'British Military Chief Reveals New Legal Fears over Iraq War', *The Observer* (1 May 2005) at <http://observer.guardian.co.uk/politics/story/0,6903,1474276,00.html>.

101   Goldsmith, 'Resolution 1441' (note 54) p. 23.

102   M. Boyce, 'Interview: Admiral Sir Michael Boyce, Conducted by Antony Barnett, April 29 2005', *The Observer* (1 May 2005) at <http://observer.guardian.co.uk/politics/story/0,6903,1474607,00.html>.

103   Ibid.

the tensions he had sought to cover and minimize between the path he was trying to tread and the US project.

It also exposes the other tension explored in this paper: that within the notion of territorial integrity, between territorial preservation and territorial sovereignty. Just as Iraq's territorial sovereignty was being violated, the preservation of its existing territorial settlement was a priority. It was underlined in the Azores Summit statement;[104] Blair noted it as a priority in his speech to the House of Commons in advance of war[105] and it appeared as an explicit war aim in briefing papers;[106] and Bush promised that 'we will provide security against those who try to spread chaos, or settle scores, or threaten the territorial integrity of Iraq'.[107] Blair similarly gained an assurance from Jalal Talabani of the Patriotic Union of Kurdistan that 'they would not try to form an independent state'.[108] The question of the territorial settlement in Iraq is obviously too large to consider here, but it is worth noting that while humanitarian reasons have been advocated for the violation of one pillar of territorial integrity (sovereignty) there seems little support outside of the theorists for the questioning of the other pillar, that of territorial preservation.[109] Indeed, for Blair this is not on the table: 'Today boundaries are virtually fixed. Governments and people know that any territorial ambition threatens stability, and instability threatens prosperity'.[110]

The relation of territorial preservation to stability and then to prosperity is revealing. James Naughtie recounts a conversation with Blair where he was asked for the relation between his thought and that of neo-conservatism's advocacy of 'the primacy of American values', the right to pre-emptive action and a dislike of internationalism. Blair suggested that, 'I come at this from a completely different perspective – a progressive perspective that says there should not be a doctrine of non-intervention in every set of circumstances. Why should the Left never support that?'[111] Between the double negative and the question lies Blair's dilemma. Why do some interventions gain legitimacy and others not? This tension remained

---

104   G. Bush, T. Blair and J.-M. Aznar, 'Azores Summit Statement' (16 March 2003) at <http://news.bbc.co.uk/2/hi/middle_east/2855567.stm>.

105   T. Blair, 'Prime Minister's Statement Opening Iraq Debate' (note X).

106   See for example, Simpson (note X) p. 12 and p. 44.

107   G. Bush, 'President Discusses the Future of Iraq, American Enterprise Institute, Washington Hilton Hotel' (26 February 2003) at <http://www.whitehouse.gov/news/releases/2003/02/20030226-11.html>.

108   Stothard (note 80) pp. 76–7.

109   For some relevant literature see, for example, Anderson and Stansfield (note 51); S. Elden, 'Reconstituting Iraq', in D. Cowen and E. Gilbert (eds), *War, Citizenship, Territory* (New York: Routledge 2008) 147–76; B. O'Leary, J. McGarry and K. Salih (eds), *The Future of Kurdistan in Iraq* (Philadelphia: University of Pennsylvania Press 2005).

110   T. Blair, 'Speech at George Bush Senior Presidential Library, Texas' (8 April 2002) at <http://politics.guardian.co.uk/speeches/story/0,11126,680866,00.html>.

111   Naughtie, *The Accidental American* (note 43) p. 72.

unresolved at the end of Blair's tenure as Prime Minister, and continues to haunt his role on the international stage today.

In Halper and Clarke's phrasing, 'neo-conservatism is not updated Reaganism. It is a new political animal born of an unlikely mating of humanitarian liberalism and brute force'.[112] Blair sacrificed the former to join the latter, but his relative strength compared to Bush did not make this bipolar or bilateral, far less international, but merely a very junior partner to exceptionalism. Blair was too far in to turn back, and became merely an ally to 'dress' the US when the UN failed to 'bless' them. On the one hand territorial integrity is sacrosanct, on the other entirely conditional. Territorial preservation is retained with as much force as can be mustered, while territorial sovereignty is rendered contingent. What is crucial is that the nation-state-territory linkage is conceived as immutable precisely in order that sovereignty can then be questioned, but that this is done without an investigation of the status of territorial settlements themselves.

---

112 Halper and Clarke, *America Alone* (note 41) p. 181.

Chapter 3

Containers of Fate: Problematic States and Paradoxical Sovereignty

Alex Jeffrey

Introduction

On 19 March 2008 the United Kingdom launched its first National Security Strategy.[1] This initiative was designed to outline an integrated foreign and domestic response to the risks facing the UK that, the document argues, have diversified and transformed since the end of the Cold War. The report articulates a range of emerging threats which it relates to the more 'complex and unpredictable'[2] security landscape following the demise of the two Cold War power blocs. In their place the strategy outlines a set of transnational and interconnected security concerns, ranging from international terrorism through to global disease pandemics. The report is consequently imbued with the language of deterritorialization and globalization: flows, interconnections and instantaneity are cited as the markers of the new security paradigm. But punctuating this apparently aspatial narrative are common references to at least one territorial dimension to this interdependent world: 'failed' or 'fragile' states. Echoing the approach taken by the National Security Strategies released by the United States in 2002 and 2006, the UK Strategy isolates 'failed states' as sites of instability and breakdown that pose a threat to the security of the UK and its allies. Such states, the report suggests, 'lack the capacity and, in some cases, the will adequately to address terrorism and organized crime, in some instances knowingly tolerating or directly sponsoring such activity'.[3] Indeed, announcing the launch of the National Security Strategy to the House of Commons UK Prime Minister Gordon Brown remarked that failed states constituted 'as great a potential threat as nuclear weapons proliferation and as demanding of a coordinated response'.[4]

---

1   See Cabinet Office, *The National Security Strategy of the United Kingdom: Security in an Interdependent World* (2008) at <http://www.cabinetoffice.gov.uk/reports/national_security_strategy.aspx>.

2   Ibid., p. 3.

3   Ibid., p. 14.

4   G. Brown, speech to the UK House of Commons setting out New Security Strategy (19 March 2008) at <http://news.bbc.co.uk/1/hi/uk_politics/7304999.stm>.

The identification of failed states as threats to the global order of state sovereignty is not a new phenomenon; the concept was applied to a number of postcolonial African states such as Nigeria, Angola and Zaire in the 1960s and 1970s.[5] But the idea of state failure has received increased policy attention in the post-Cold War and, more particularly, the war on terror era, reflected in the enrolment of the term in the very heart of UK and US security strategies. As indicated in the UK Strategy, current declarations of state failure constitute part of a wider practice of connecting adjudications of the capabilities of states with global security concerns. In fact, the categorization of statehood that emerges from UK and US security discourse sketches a spectrum of threat to the global order, from fragile and weak states (potential threat), through failed states (probable threat) to rogue states (likely or imminent threat).

This typology has emerged out of more formal analyses in development studies, international relations and political science and the practical rationalizations of policymakers.[6] Within this synthetic discourse we find the idea that states that are problematic for global order constitute a distinct governance problem requiring a set of technical solutions derived from expert analysis. Within this framework there is often particular concern over supra- and sub-state networks and processes, in particular the effects of criminal, nationalist, militia or terrorist groups challenging the status quo by 'exploiting the territorial vacuum of power'.[7]

A second more critical perspective eschews the technical aspects of what constitutes a failed or rogue state and focuses instead on the political, historical and geographical imaginaries that underpin adjudications of state sovereignty. This approach is reflected in the work of scholars such as Duffield,[8] Chomsky,[9] and Bilgin and Morton,[10] who have attempted to trace the moral assumptions and political

------

5 Writing in 1982, Jackson and Rosberg note 'there have been times when Angola, Chad, Nigeria, Sudan, Ethiopia, Uganda and Zaire have ceased to be "states" in the empirical sense – that is their central governments lost control of important areas of their jurisdiction during struggles with rival political organisations'. See R. Jackson and C. Rosberg, 'Why Africa's Weak States Persist: The Empirical and the Juridical in Statehood', *World Politics* 35/1 (1982) 1–24, p. 1.

6 See, for example, J. Milliken and K. Krause, 'State Failure, State Collapse, and State Reconstruction: Concepts Lessons and Strategies', *Development and Change* 33/5 (2002) 753–74; M. Rotberg (ed.), *When States Fail: Causes and Consequences* (Princeton and Oxford: Princeton University Press 2004).

7 A. Yannis, 'State Collapse and its Implications for Peace-building and Reconstruction', *Development and Change* 33/5 (2002) 817–35, p. 818.

8 M. Duffield, *Global Governance and New Wars* (London and New York: Zed Books 2001); M. Duffield, 'Development, Territories, and the People: Consolidating the External Sovereign Frontier', *Alternatives* 32/2 (2007) 225–46.

9 N. Chomsky, *Failed States: The Abuse of Power and the Assault on Democracy* (London: Penguin Books 2006).

10 P. Bilgin and D. Morton, 'Historicising the Representation of "Failed States": Beyond the Cold War Annexation of the Social Sciences', *Third World Quarterly* 23/1 (2002) 55–80.

interests at work in public declarations of state failure. In particular, three points may be extracted from this critical scholarship. Firstly, the designation of states as problematic suggests a set of analytical criteria against which their competence or orientation can be objectively assessed. Secondly, such declarations suggest that Western policy makers and commentators have a privileged vista from which to make such objective pronouncements. Thirdly, the concept of state failure maps the causes of instability and institutional collapse within the boundaries of the state in question, thereby severing the wider historical and geographical networks within which every state is embedded. Within this optic, the labelling of a state in a particular way serves to absolve the adjudicating actor of any responsibility for the condition. In short, the spatialization of danger encapsulated in such adjudications is a geopolitical move: ethnocentric, disembodied and realist in its imagination of inter-state affairs.

In this chapter I will seek to build on this critique by exploring the political effects of declarations of 'failed' and 'rogue' states in the 'war on terror', drawing in particular on the case of the justification for the 2003 military intervention in Iraq. In doing so, I am not attempting to provide a definitive assessment of the characteristics of failed states.[11] Instead, I seek to explore the forms of intervention and political stances that are *enabled* by adjudications of state competence and how these practices can be understood as part of the broader geopolitics of security promoted by the 'war on terror'. In particular I am keen to situate such iterative practices within the wider historical context of neoliberal (and later neoconservative) development and security paradigms. As I outline below, declarations of state failure include both biopolitical and geopolitical components, which operate in tension. Biopolitically the labels of 'failed' and 'rogue' states draw our attention to a compulsion to intervene to save human life on account of the failings of the nominally sovereign power to protect its citizenry, while geopolitically they territorialize deviance and culpability within the borders of the states in question. I argue that this dialectic between the biopolitical and the geopolitical constructs the space for powerful state actors to advance a development agenda founded upon neoconservative and neoliberal ideals.

In making this argument I wish to advance Bilgin and Morton's claim that labels such as 'failed' and 'rogue' states constitute 'representations that enable certain policies which serve the economic, political and security interests of those who employ them'.[12] In examining this productive nature of discourses of state failure I am drawing on the theoretical insights of scholars of critical geopolitics, a body of research that has drawn on poststructural theorists (notably Michel Foucault and Jacques Derrida) to illuminate the politics of writing global

---

11   For such an undertaking see Rotberg (note 6).
12   Bilgin and Morton (note 10) p. 56.

space.[13] Rather than accepting geopolitical pronouncements as reflections of an external reality, scholars of critical geopolitics have explored the functioning of geographical knowledge 'as an ensemble of technologies of power concerned with the governmental production and management of territorial space'.[14] Within this framework geopolitical writing is identified as a 'discursive practice, by which intellectuals of statecraft "spatialize" international politics and represent it as a "world" characterized by particular types of places, peoples and dramas'.[15] Therefore this work draws attention to the productive capacity of geopolitical analysis, specifically the mechanisms through which it 'naturalizes and normalizes particular political claims as "geographical" and hence given'.[16] I argue that discourses of state failure are particularly open to critical deconstruction. Such an approach focuses our attention on the situated and partial nature of adjudications of state failure, highlighting that they are always iterations that emerge from distinct political perspectives rather than reflecting disembodied empirical assessments. Therefore I suggest that assessments of the adjudication of Iraqi statehood should not focus on the content of such pronouncements but rather their political effects. The identification of Iraq in 2003 as both a failed and rogue state involved an attempt to territorialize deviance within Iraq's borders and sever the longer historical lineage of UK and US engagement in Iraq in order to produce a coherent object of intervention.

I make this argument across two sections. The first examines the conceptual position of the state within current Western security strategies. This narrative explores the paradoxical spatialization of sovereignty within current Western foreign policies: namely, the right to adjudicate and intervene in global affairs whilst simultaneously identifying the state as the pre-eminent container of political power. Hence state sovereignty is both contingent upon particular political, economic or moral conditions and upheld as the indivisible territorialization of politics. This discussion draws attention to the historical production of paradoxical sovereignty, in particular through the fusing of security and development concerns within the policies of Western states in the late 1990s. This narrative focuses attention on the politicization of development policy, citing what Mark Duffield[17] has identified as the biopolitical emphasis of recent development/security initiatives, where interventions have targeted the protection of human populations rather than the development of individual states. Within this template, self-identified 'responsible

---

13   G. Ó Tuathail, *Critical Geopolitics* (London: Routledge 1996); G. Ó Tuathail and J. Agnew, 'Geopolitics and Discourse: Practical Geopolitical Reasoning in American Foreign Policy', *Political Geography* 11/2 (1992) 190–204.

14   Ó Tuathail (note 13) p. 7.

15   Ó Tuathail and Agnew (note 13) p. 190.

16   M. Kuus, *Geopolitics Reframed: Security and Identity in Europe's Eastern Enlargement* (New York and Houndmills: Palgrave Macmilan 2007) p. 7.

17   M. Duffield, 'Getting Savages to Fight Barbarians: Development, Security and the Colonial Present', *Conflict, Security and Development* 5/2 (2005) 141–59.

states' (the conceptual obverse of problematic states) cite humanitarian causes to suspend the sovereignty of states that are deemed to be failing to protect their citizens or are actively oppressing them. This focus on human security draws our attention to how international institutions 'categorise, separate and act upon'[18] populations in the developing world. Utilizing Bilgin and Morton's concept of 'embedded statism',[19] I argue that this process of categorization and separation has served to entrench, rather than replace, an enduring attachment to states as territorial 'containers of power'[20] within dominant development and security discourses. From a political economy perspective, states provide certain capacities and services required by neoliberal development, such as regulation of international capital and the main institutions of democratic practice. From a geopolitical perspective, the state comprises a fixed and bounded territory onto which particular competences may be projected. Contemporary adjudications of statehood are therefore preoccupied with increasingly globalized flows, migrations and mobilities on the one hand, and the competence and responsibility of individual states and their territories on the other, but confine their understanding of these within narrow and self-serving channels. Furthermore, the ambiguities within this conceptual space afford considerable flexibility for policy action.

The second section offers an examination of UK and US geopolitical scripting of the 2003 military interventions in Iraq. Opening with a discussion of the cases of Afghanistan and Pakistan I explore the differentiated vocabularies of problematic statehood. As Simon Dalby has argued, the mapping of deviance onto the Afghan state fixed the terrorist threat onto a specific territory and served to justify the subsequent military intervention 2001–02.[21] In the case of Iraq, the Bush and Blair governments labelled the state as both 'failed' and 'rogue', citations that shaped the institutional form and stated objectives of the subsequent intervention. I argue that this scripting of Iraq encompassed three interconnected spatial and temporal manoeuvres, reflecting the varied and shifting justifications for the intervention. Firstly, UK and US foreign policy pronouncements identified the regime of Saddam Hussein as a global threat. Rather than confined to its own territory and population, which had previously been subject to state violence, Iraq's weapons of mass destruction (WMD) were cited in numerous policy documents as an imminent threat to world peace. Secondly, in the wake of the failure to find any evidence of WMD, policy makers presented the intervention in biopolitical terms, a morally just act to confront the ongoing humanitarian suffering of the Iraqi population. Finally, attempts were made to project responsibility for containing violence subsequent to the invasion onto the new Iraqi state. Colonial, Cold War

---

18   Ibid., p. 2.

19   Bilgin and Morton (note 10) p. 58.

20   See P. Taylor, 'The State as Container: Territoriality in the Modern State System', *Progress in Human Geography* 18/2 (1994) 151–162.

21   S. Dalby, 'Calling 911: Geopolitics, Security and America's New War', *Geopolitics* 8/3 (2003) 61–68.

and more recent connections with the Saddam Hussein regime were erased in policy documentation and discursive performances; in their place deviance was territorialized within the borders of Iraq. These spatial constructions were key to the justifications offered for military intervention. This spatial imaginary also set the parameters for a potential solution: the construction of a state in the images of neoliberalism and neoconservatism in the post-invasion period. Within this framework the conflict is justified in state-territorial terms, whereby a failing and rogue state is to be transformed into a beacon of democracy in the Middle East.

In conclusion I explore the implications of the rhetoric of state failure for current attempts to foster democracy through external intervention. Such practices are enabled by a technical understanding of statehood, that there exists a set of remedial practices that can establish neoliberal and democratic structures in contexts that previously failed to meet the desired template. The morality of recent interventions within the 'war on terror' has rested on such perceived corrective virtues. The conclusion questions such narratives. Just as declarations of 'state failure' have depended on an enduring belief in the capacity for state boundaries to delineate deviance, discourses of 'state building' have relied on the image of coherent state sovereignty despite intrusive international intervention and the reality of fragmented control and breakdown. Through the embedded statism of development and security discourse, post-intervention claims of democratic sovereignty map the failures of intervention on the intransigence and incapacity of a failed state's population and evade the responsibility of the intervening agencies for the consequences of their actions.

Embedded Statism

The end of the Cold War marked a prolonged period of geopolitical self–assessment by Western states. However, in the case of the US, Ó Tuathail notes that 'rather than the end of the Cold War leading to a critical reassessment of the national security agencies and intellectuals that had built the Soviet Union into such an awesome threat, bureaucratic and ideological relegitimation were the order of the day'.[22] This task of relegitimation required a rethinking of the established geopolitical categories of Self and Other, moving away from the binary division of the world that had provided a sense of stability and certainty in mapping threat over the previous forty years. In its place, security discourses began to evoke 'uncertainty' and 'unpredictability' as threats posed in the new post-Cold War world order.[23] These new security discourses were structured around the ubiquity of neoliberal globalization and the existence of a single superpower. As Bialasiewicz and Elden

---

   22   G. Ó Tuathail, 'Introduction to Part Two', in G. Ó Tuathail, S. Dalby and P. Routledge, *The Geopolitics Reader* (London: Routledge 2006) 59–73, p. 71.
   23   G. Ó Tuathail, 'The Bush Administration and the "End" of the Cold War: A Critical Geopolitics of US Foreign Policy in 1989', *Geoforum* 23/4 (1992) 437–52.

suggest, the practice of rethinking security threats also entailed a process of reterritorialization, or 'the remaking of spatial and geographical relations rather than the deterritorialisation that globalisation is often supposed to be'.[24]

Two interlinked points need to be made regarding the new reterritorialized world order. First, reterritorialization reproduced what Bilgin and Morton term the embedded statism within powerful institutions of knowledge production (such as universities, think tanks or policy units). This term refers to the unquestioned and naturalized reliance on the state as the primary container of political life. They argued that the Cold War annexation of the social sciences in the US has limited communication between politics and economics, masking the socioeconomic history of the emergence of current security concerns surrounding failed states. Second, the reterritorialization of security threats has been structured around a fusion of ideas of development and security. As Duffield notes, 'through a circular form of reinforcement and mutuality, achieving one is regarded as essential for securing the other'.[25] This security paradigm positioned underdevelopment as a threat to developed states through the spread of instability and therefore required intervention to mediate the possibility of the recurrence of violent conflict. Duffield has furthermore identified this development–security nexus as a biopolitical configuration, arguing that while the main concern of development interventions has orientated towards human populations rather than the states in which they live, the purpose of development is still conceived as being to maintain life in a form that is supportive of global order, rather than to achieve emancipation through political transformation.

The principles of embedded statism and the development–security nexus are evident in the recently produced security strategies by both the UK and the US. For example the 2006 US National Security Strategy promotes the establishment of a tapestry of sovereign states:

> In the world today, the fundamental character of regimes matters as much as the distribution of power among them. The goal of our statecraft is to help create a world of democratic, well-governed states that can meet the needs of their citizens and conduct themselves responsibly in the international system. This is the best way to provide enduring security for the American people.[26]

As reflected in this statement, the US and UK security strategies published since the launching of the war on terror have argued that there exists a single

---

24   L. Bialasiewicz and S. Elden, 'The New Geopolitics of Division and the Problem of a Kantian Europe', *Review of International Studies* 32/4 (2006) 623–44, p. 627. See also S. Elden, 'Missing the Point: Globalisation, Deterritorialisation and the Space of the World', *Transactions of the Institute of British Geographers* 30/1 (2004) 8–19.

25   Duffield, *Global Governance and the New Wars* (note 8) p. 16.

26   The White House, *The National Security Strategy of the United States of America* (2006) at <http://www.whitehouse.gov/nsc/nss/2006/> p. 1.

model for global security founded upon the intersection of neoliberal economic theory, democratic governance and the sanctity of the state as the primary unit of international relations. There is nothing new in this discursive connection; as Bilgin and Morton note, development interventions since US President Truman's inaugural address in 1949 have been directed towards fostering economic and social progress in 'underdeveloped countries'.[27] Thus, in the war on terror, 'development', 'good governance', 'humanitarian intervention' and 'regime change' are all required to prevent fragile states becoming failed states, to build state functions where they are absent, and to bring rogue states back within the fold of the international community. However, to understand the ambiguous placement of state sovereignty within contemporary security/development strategy requires an historical analysis of the primacy given to neoliberal economic ideology and the enrolment of democratic rhetoric within development interventions.

The privileged position of neoliberalism within development interventions can be traced to the slow (and by no means absolute) replacement of Keynesianism with a neoliberal political economic doctrine in the administrations of the US and Western Europe over the course of the 1970s. David Harvey[28] sketches a series of paradigmatic cases in this capitalist reorganization of state functions, from the corporate solution to the New York fiscal crisis 1975–76 to the breaking of the 1984 miners' strike in the UK under Margaret Thatcher.[29] Harvey uses these moments (among others) to highlight the shift in state priorities, from ensuring the wellbeing of society through social welfare functions to providing a regulatory framework for the circulation and accumulation of private capital. In line with this prioritization, the 1980s saw the Reagan and Thatcher administrations dismantle state monopolies over key industries in the US and UK, a policy presented as an economic and moral imperative.[30] In addition, the spaces exposed by the rollback of state apparatus provided opportunities for a range of 'non-governmental organizations' (NGOs) to fulfil welfare roles.[31] A number of scholars have

---

27   Bilgin and Morton (note 10) p. 59. See also A. Escobar, *Encountering Development* (Princeton: Princeton University Press 1995).

28   D. Harvey, *A Brief History of Neoliberalism* (Oxford: Oxford University Press 2005).

29   See also N. Klein, *The Shock Doctrine The Rise of Disaster Capitalism* (London: Allen Lane 2007); J. Peck and A. Tickell, 'Neoliberalizing Space', *Antipode* 34/3 (2002) 380–404.

30   D. Swann, *The Retreat of the State: Deregulation and Privatisation in the UK and US* (Brighton: Wheatsheaf 1988); C. Jones, 'A Regional Perspective on the Impact of the Privatisation of the UK Public Industrial Stock', *Environment and Planning C* 23/1 (2005) 123–39.

31   K.-L. Tang, 'The Case for the Privatization of Social Welfare: Three Decades of Public Opinion Evidence from Great Britain and the United States', *The Scandinavian Journal of Social Welfare* 6/1 (1997) 34–43.

characterized this shift as a move from government (state centred) to governance (decentred and multi-institutional).[32]

Though the rise of neoliberalism reshaped the internal practices of governance within wealthy states, the common interchanging of neoliberalization and globalization speaks of the conceptual difficulty of restricting this process to individual states. In particular, neoliberalism was dependent upon the internationalization of finance and the opening of new spaces of production, accumulation and exchange. A crucial moment in the export of this doctrine from wealthy states to the developing world came in the early 1980s with the Reagan administration's harnessing of the International Monetary Fund (IMF) to police the reorganization of states in the developing world down neoliberal lines.[33] This opportunity emerged out of the debt crises experienced across the developing world over the 1980s. Interest rate increases on debts granted through the recycling of petrodollars following the 1970s oil price hikes left many Latin American and Sub-Saharan African countries on the verge of defaulting to public and private creditors in US and Western Europe. This financial disequilibrium provided the necessary political leverage to enforce IMF demands coalescing around the reform of state architectures to conform to established neoliberal norms. Often referred to as 'austerity measures' or 'shock therapy', the IMF structural adjustment programmes led to greater social inequalities as state service provision was reduced in favour of market-led reform.[34]

The reconfiguration of state competencies under conditions of 'structural adjustment' has been portrayed in certain quarters as an opportunity for democratic participation and popular engagement in processes of governance.[35] The US National Security Strategy of 2006 appears to agree, arguing that globalization has 'helped the advance of democracy by extending the marketplace of ideas and the ideals of liberty'.[36] As in the case of the earlier neoliberalization of wealthy countries, the retreat of the state from social functions in the developing world coincided with an increase in NGOs directed towards meeting welfare needs.[37] Donor and inter-governmental organizations cast the plurality of associations fulfilling social

32   B. Jessop, 'Capitalism and its Future: Remarks on Regulation, Government and Governance', *Review of International Political Economy* 4/3 (1997) 561–607.

33   W. Bello, S. Cunningham and B. Rau, *Dark Victory: The United States, Structural Adjustment and Global Poverty* (London: Pluto Press 1998).

34   See for example U. Cizre and E. Yeldan, 'The Turkish Encounter with Neoliberalism: Economics and Politics in the 2000/2001 Crises', *Review of International Political Economy* 12/3 (2005) 387–408; J. Petras and H. Brill, 'The IMF, Austerity and the State in Latin America', *Third World Quarterly* 8/2 (1986) 425–48.

35   D. Ghai, *Structural Adjustment, Global integration and Social Democracy* (Geneva: The United Nations Research Institute for Social Development 1992).

36   The White House (note 26) p. 47.

37   L. Salamon, 'The Rise of the Non-profit Sector', *Foreign Affairs* 73/4 (1994) 109–22.

welfare objectives as evidence of 'good governance',[38] a policy discourse that advanced the neoliberal values of 'transparency' and 'accountability' beyond the management of the economy and into wider political life.[39] Though laudable, these objectives do not neatly equate to representative democracy. Most crucially, 'good governance' 'does not prejudge the locus of actual decisionmaking, which could be within the state, within an international organization or within some other structural context'.[40] Discourses of 'good governance' can consequently be characterized as advocating a normative 'neoliberal' democratic doctrine, replacing majority rule with the construction of 'correct institutions', 'transparent decision making' and 'associational pluralism'. These processes seem to provide tentative support to Harvey's observation that neoliberalism tends 'to favour governance by experts and elites' rather than majority rule.[41]

Despite the disjuncture between objectives of neoliberal 'good governance' and democratic decisionmaking, the democratization of Latin America in the 1980s and the end of the Cold War in 1991 offered a new opportunity to celebrate democratic neoliberalism. The fall of authoritarian regimes in Eastern and Central Europe and the Soviet Union marked the beginning of their transition to democracy and neoliberalism. Francis Fukuyama infamously described this moment as the 'end of history', a celebratory refrain for the 'victory' of democratic neoliberalism and the associated foreclosure of alternative modes of political economic organization.[42] This account, however, paid little attention to the profoundly undemocratic nature of the neoliberalization of the developing world, a process of economic coercion conducted through undemocratic institutions (not least the IMF and World Bank) and often authoritarian regimes (General Pinochet's Chile, for example). In light of this uneasy historical relationship between neoliberalization and democratization, their unproblematic combination after the Cold War was a shrewd discursive manoeuvre on the part of their champions in the US and Europe. Accounts of 'transition' from Communist pasts to market-led democratic futures in post-socialist Europe often blurred practices of democratization with neoliberalization, each drawing on the same language of 'transparency' and 'accountability'.

Despite the increasing globalization of neoliberal doctrine, and associated claims of 'deterritorialization', recent development and security discourses within the UK and US have continued to promote the indivisibility of state sovereignty.[43] The continued state-centred focus of practices of intervention seems, at first blush,

---

38   World Bank, *Sub-Saharan Africa: From Crisis to Sustainable Growth* (Washington DC: World Bank 1989).

39   A. Leftwich, 'Governance, Democracy and the Development of the Third World', in S. Corbridge (ed.), *Development Studies: A Reader* (London: Arnold 1995) pp. 427–38.

40   M. Doornbos, '"Good governance": The Rise and Decline of a Policy Metaphor', *Journal of Development Studies* 376 (2001) 93–108.

41   Harvey (note 28) p. 66.

42   F. Fukuyama, *The End of History and the Last Man* (London: Penguin 1992).

43   See S. Elden, 'Territory and Terror', *Antipode* 39/5 (2007) 821–45.

paradoxical when considered in light of the neoliberal rhetoric of individualism, free enterprise and the retreat of the state. But the image of the world as divided into a tapestry of sovereign states has proved vital for the policing of neoliberalism and retaining a preferred institutionalization of democracy. This focus on institutionalization and regulation highlights a shift in the 1990s from neoliberal thinking (based on free market and individualistic concerns) to more neoconservative approaches (highlighting the need for order and codes of morality within the global system).[44] But the neoconservative perspective has enhanced, rather than undermined, the embedded statism of security and development concerns. Within neoconservative thought the state provides a series of prerequisites for economic expansion (a system of property rights, law, policing and defence, and the supply of money and public goods) that no individual would be able to supply at a profit.[45] The primacy of states as the territorial expression of political community also defers democratic reform away from transnational or inter-governmental agencies or processes. Reflecting the conflation of development and security objectives, scholars and policymakers focused on the necessity to establish democratic states as a means of mediating interstate conflict on a global scale.[46] This 'democratic peace' model relies on the premise that no two democracies had ever been in conflict, hence a globe comprising solely of democratic states would, it is asserted, usher in an era of Kantian 'perpetual peace'.[47] Thus, the establishment and reinforcement of a globe constructed of a tapestry of sovereign and democratic states has gained a security imperative as a means of managing international threats.

Recent US and UK security strategy has echoed the virtuous connections between democracy and state sovereignty found within democratic peace theories. In the US case this should perhaps come as no surprise. The founders of the Project for a New American Century (PNAC – a neoconservative 'education organization' with roots in the 1970s) had since the 1990s advocated the 'expansion of the zone of democratic peace' through American military intervention in the Middle East, and some of its members went on to hold prominent positions within the White House (Dick Cheney) and Pentagon (Paul Wolfowitz). But as Bialasiewicz and Elden have identified, it is imprecise to describe US security strategy as solely

---

44   Harvey (note 28) p. 82. For an extended account of the distinctions between neoliberal and neoconservative thought see J. Glassman, 'The *New* Imperialism? On Continuity and Change in US Foreign Policy', *Environment and Planning A* 37/9 (2005) 1527–44.

45   See D. Harvey, 'The Marxian Theory of the State', *Antipode* 8/2 (1976) 80–89; J. Painter, 'State–society', in P. Cloke and R. Johnston (eds), *Spaces of Geographical Thought: Deconstructing Geography's Binaries* (London: Sage 2005) pp. 42–60.

46   A. Franceschet, 'Popular Sovereignty or Cosmopolitan Democracy? Liberalism, Kant and International Reform', *European Journal of International Relations* 6/2 (2000) 277–302.

47   C. Layne, 'Kant or Cant? The Myth of Democratic Peace', *International Security* 19/2 (1999) 5–49.

rooted in a Kantian imaginary of a model of democratic peace.[48] In critiquing the recent interventions by high profile intellectuals of statecraft such as Robert Kagan and Thomas Barnett, Bialasiewicz and Elden rightly point to the uneasy position of state sovereignty within such narratives: state sovereignty must be enshrined as indivisible, but simultaneously may be transgressed by privileged actors (the US) in order to correct institutional failings. As they note, this position seems to encapsulate a combination of Hobbesian realism, the need for a strong state to act unilaterally, and Kantian idealism, the desire to move towards a world of democratic and peaceful states. It is not the place of this chapter to engage in an exploration of these competing political philosophies (see Chapter 2), but they are an important underpinning to the reliance of US security discourses on a 'realist' narrative of international affairs, as sketched in the discussions above, where security threats are posed by weakening state capacity and undemocratic governance by unsuitable dictators. As Duffield and Waddell note, '[t]he perception of these rulers as the illegitimate enemies of development, together with concerns that disaffected people are liable to be drawn to them, establishes an interventionalist dynamic'.[49]

Considering the origins of the US National Security Strategies as responses to the terrorist attacks of September 11 2001, it is perhaps no surprise that they offer a Manichean vision of a world divided between 'forces of freedom' (the US and its allies) and the 'enemies of civilization': the 'terrorists'.[50] This geopolitical imaginary reworks the Cold War narrative of a virtuous capitalist Self in opposition to a communist Other, replacing it with a more open–ended binary of freedom–loving Self against a more plural and diffuse terrorist Other. But the crucial aspect of these abstracted geopolitical imaginaries is their reattachment to specific territories through the vocabulary of 'failed' and 'rogue' states. Threat has been reterritorialized onto the imagined deviant states that fail to meet prescribed criteria in terms of governance and sovereignty. Therefore the adjudication of state sovereignty is not merely a description of state capacity but a profoundly productive labelling act. It serves to disconnect the US and the UK (and their allies) from any historical or contemporary complicity in social, economic and political breakdown whilst presenting themselves as well placed to enact a remedial humanitarian intervention. I will go on to explore these practices in the case of US-led intervention in Iraq.

---

48   Bialasiewicz and Elden (note 24), see also, S. Roberts, A. Secor and M. Sparke, 'Neoliberal Geopolitics', *Antipode* 25/5 (2003) 886–97.

49   M. Duffield and N. Waddell, 'Securing Humans in a Dangerous World', *International Politics* 43/1 (2006) 1–23, p. 7.

50   See The White House, *The National Security Strategy for the United States of America* (2002) at <http://www.whitehouse.gov/nsc/nss.html> pp. 5–7.

Problematic States and the War on Terror

The Bush administration's launch of a 'war on terror' following the attacks of 11 September 2001 acted as a mechanism for the justification of military intervention in a number of sovereign states (see also Chapter 5). The abstract and deterritorialized nature of terrorism as an adversary ensured that this was never going to be a straightforward interstate conflict. However, an initial priority for the US and its allies was the identification of a clear territorial target for the envisaged military action. This is most acutely reflected in US Vice President Dick Cheney's comments at the emergency National Security Council meeting on 11 September 2001: 'To the extent that we define our task broadly, including those who support terrorism, then we get at states. And it's easier to find them than Bin Laden.'[51] While exhibiting the enduring embedded statism of US government discourse, this approach also reflects an acute performance of the development–security nexus. The interventions in the 'war on terror' have entangled militaristic and humanitarian vocabularies to convey a biopolitical concern for the welfare of target populations while simultaneously threatening military action on target governments. As a number of scholars of geopolitics have noted, the simple speech act of connecting 'terror' with specific states, as in the case of the imaginary of the 'axis of evil' conjured in the 2002 State of the Union address, helped bolster US public opinion in favour of military intervention, harnessing the symbolism of the terrorist attacks to legitimize military action and the domestic loss of civil liberties.[52]

The first target set out by the Bush administration was Afghanistan, a state that was reported by the US President to be harbouring terrorists, and thus had 'failed' in its ability to govern its territory.[53] This criterion of failure could have been applied to other states, most notably neighbouring Pakistan, which was an ally of the Taleban and had, since 1999 been ruled by its military leader General Pervez Musharraf. But reflecting the flexible nature of 'failed state' labelling, President Bush presented Pakistan as a site of strength and competence, describing the state as 'a strong ally' and President Musharraf as 'a strong leader' for which 'the world is deeply appreciative'.[54] Pakistan's support was crucial for logistical and symbolic reasons, in particular supporting President Bush's assertion that the conflict in Afghanistan was not a strike on the Islamic faith but rather a targeted strike on specific terrorist groups and the provision of humanitarian relief.[55] As Derek

51   D. Cheney, cited in J. Kampfner, *Blair's Wars* (London: The Free Press 2003) p. 156.

52   See, for example, K. Dodds, *Geopolitics: A Very Short Introduction* (Oxford: Oxford University Press 2007).

53   See G. Bush, *Address to Joint Session of Congress and the American People* (2001) at <http://www.whitehouse.gov/news/releases/2001/09/20010920-8.html>.

54   G. Bush, *President of Pakistan Reaffirms Commitment to Fight Terrorism* (2001) at <http://www.whitehouse.gov/news/releases/2001/11/20011110-6.html>.

55   The consent of Pakistan to assist in the military action in Afghanistan had been established through the assurance of debt relief, resumption of diplomatic relations and

Gregory[56] and Simon Dalby[57] have suggested, the connection of Afghanistan and terror required two interlinked spatial manoeuvres. The first was a 'performance of sovereignty' whereby the fragmented space of Afghanistan could be represented as a coherent state. The second was a 'performance of territory' through which 'the fluid networks of al-Qaeda could be fixed to a bounded space'.[58]

But this geopolitical framing of Afghanistan was set within a broader biopolitical rationale for the military intervention. The mapping of terrorism onto the Afghan state was accepted by the UN Security Council who, in passing Resolution 1378, suggested they were 'deeply concerned by the grave *humanitarian situation* and the continuing serious violations by the Taliban of human rights and international humanitarian law'.[59] The humanitarian justification for the intervention was further emphasized in former UK Prime Minister Tony Blair's speeches prior to the military action, in one case suggesting 'the conflict will not be the end. We will not walk away, as the outside world has done so many times before'.[60] In addition, Cynthia Weber notes how President Bush urged US citizens to watch Mohsen Makhmalbaf's *Kandahar*, a 2001 film that charts the journey of an Afghan-born Canadian woman as she travels to Afghanistan to find her sister.[61] The film was made before the attacks of September 2001 and offers an account of the oppression of women within Taliban-controlled Afghanistan. Bush therefore drew on what could be termed the film's popular biopolitics – it conveyed the threat posed to women under the Taliban regime and the resultant opportunity for female liberation offered by US-led military intervention.[62] The Afghanistan intervention consequently demonstrates the ability of the failed state marker to both spatialize the development–security nexus while simultaneously providing a recognizable target for military intervention.

The intervention in Afghanistan, however, marked only the opening salvo on a broader military campaign envisaged by the US within the 'war on terror'. As we have seen, the projection of threat was widened to more abstract concepts of failing state sovereignty and poor governance. As I have discussed, this prescription of threat was not restricted to official US security strategies and pronouncements. The orientalized imagery of an 'undemocratic' and 'untamed' Other was reproduced in

---

the prospect of the solidification of the Northern border of Pakistan with Afghanistan. See Kampfner (note 51) pp. 120–126.

56 D. Gregory, *The Colonial Present* (Oxford: Blackwell 2004) pp. 49–50.

57 Dalby, 'Calling 911' (note 21) pp. 72–3.

58 Elden, 'Territory and Terror' (note 43) p. 824.

59 United Nations Security Council, *Resolution 1378* (New York, United Nations 2001) at <http://www.state.gov/s/ct/rls/other/6138.htm>. Emphasis added.

60 Kampfner (note 51) p. 122.

61 C. Weber, 'Not Without My Sister: Imagining a Moral America in "Kandahar"', *Open Democracy* (11 September 2005) at <http://www.opendemocracy.net/democracy-resolution_1325/kandahar_3006.jsp>.

62 I would like to thank Klaus Dodds for drawing my attention to this example.

a series of neoconservative policies and publications.[63] The geographical fulcrum of these imaginaries was the Middle East, and most specifically, Iraq. The Bush administration had sought a break from the Clinton policy of containment (albeit with continued bombing of military and 'dual-use' targets) and the 'war on terror' provided a vehicle of justification for regime change.

There has been a wealth of recent critical scholarship examining the ideological, economic and symbolic underpinnings to the Bush administration's preoccupation with the Iraqi invasion. For some, the invasion can only be understood through the lens of political economy, in particular the US desire to gain control over Iraq's plentiful oil supplies and the positive economic spinoff of mass military deployment and lucrative reconstruction contracts.[64] Others have pointed to the realm of the spectacle: Iraqi invasion would work to assuage the US 'image defeat' of 11 September 2001 and act as a demonstration of US military superiority for the surrounding Middle East.[65] Leaders in the US and UK derided these geoeconomic and militaristic reasons as 'conspiracy theories', arguing instead that intervention was necessary on account of Iraq's possession of WMD, its terrorist links with al-Qaeda and its failure to protect its citizenry.[66]

In the official justifications for intervention Iraq was not simply a 'failed state', though it exhibited similar shortcomings in terms of state capacity. Iraq was also labelled a 'rogue state': armed, dangerous and acting outside of the norms of international law. In the following excerpt from a press statement marking the release of the 2002 National Strategy to Combat Weapons of Mass Destruction, President Bush explains the nature of the rogue state label:

> Some rogue states, including several that support terrorism, already possess WMD and are seeking even greater capabilities, as tools of coercion. For them, these are weapons of choice intended to deter us from responding to their aggression against our friends in vital regions of interest. For terrorists, WMD would provide the ability to kill large numbers of our people without warning. They would give them the power to murder without conscience on a scale to match their hatred for our country and our values.[67]

---

63   T. Barnett, *The Pentagon's New Map: War and Peace in the Twenty First Century* (London: Putnam 2004); R. Kagan, *Of Paradise and Power: America and Europe in the New World Order* (New York: Atlantic Books 2004).

64   See D. Harvey, *The New Imperialism* (Oxford: Oxford University Press 2003); N. Smith, *The Endgame of Globalisation* (New York: Routledge 2005).

65   Retort, *Afflicted Powers: Capital and Spectacle in a New Age of War* (London: Verso 2005), drawing on G. Debord, *The Society of the Spectacle* (London: Rebel Press 1992).

66   M. Tempest, 'Blair: Oil Claim is "Conspiracy Theory"', *The Guardian* (15 January 2003) at <http://www.guardian.co.uk/politics/2003/jan/15/foreignpolicy.uk>.

67   G. Bush, *Statement by the President* (11 December 2002) at <http://www. whitehouse.gov/news/releases/2002/12/20021211-8.html>.

Echoing the mapping of terrorism onto Afghanistan, in this excerpt President Bush evokes the dual sovereign capacity of 'rogue states'. In the first case the rogue state is perceived to exhibit an excess of sovereign power which corresponds to an ability to monopolize violence both within their sovereign borders and against 'our friends in vital regions of interest'. The statement also gestures towards the potential for rogue states and terrorist networks to come into alliance. Furthermore, a rogue state that had developed WMD but which subsequently 'failed' could have its military architecture captured and utilized by terrorists 'to murder without conscience'. The cumulative effect of these conflicting invocations of state sovereignty was a blurring of the distinction between Saddam Hussein's Iraqi regime and the terrorist threat posed by al-Qaeda. Reflecting once more the entanglement of development and security, the threat posed by the continuation or collapse of Iraqi state capacity was not restricted to its own borders, but rather rescaled to the entire world. This global scale threat was most infamously invoked in Tony Blair's introduction to the UK Government's 2002 assessment of Iraq's weapons capability:

> Saddam has used chemical weapons, not only against an enemy state, but against his own people. Intelligence reports make clear that he sees the building up of his WMD capability, and the belief overseas that he would use these weapons, as vital to his strategic interests, and in particular his goal of regional domination. And the document discloses that his military planning allows for some of the WMD to be ready within 45 minutes of an order to use them.[68]

In internationalizing the threat posed by the Hussein regime the US and UK administrations also made attempts to inspire a multilateral response. This strategy involved the tabling of a draft UN Security Council Resolution on 7 March 2003 recognizing 'the threat Iraq's non-compliance with Council resolutions and proliferation of weapons of mass destruction and long-range missiles poses to international peace and security'.[69] Despite intense lobbying, particularly by Prime Minister Blair, the UN Security Council refused to back the draft resolution.[70] This international discord served to narrow the breadth of the coalition undertaking the intervention, and consequently the institutions responsible for reconstructing Iraq in its aftermath. Understanding the symbolic capital of multilateral vocabulary, the Bush administration was keen to advertise that the war was officially prosecuted

---

68   T. Blair, 'Foreword by the Prime Minister', in *Iraq's Weapons of Mass Destruction: The Assessment of the British Government* (2002) at <http://www.number-10.gov.uk/output/Page271.asp>.

69   United Nations Security Council, *Spain, United Kingdom of Great Britain and Northern Ireland and the United States of America: Draft Resolution* (2003) at <http://www.casi.org.uk/info/undocs/scres/2003/20030307draft.pdf>.

70   See A. Danchev, 'Tony Blair's Vietnam: The Iraq War and the "Special Relationship" in Historical Perspective', *Review of International Studies* 33/2 (2007) 189–203.

by a 'Coalition of the Willing', though in real troop numbers this amounted to a US invasion (committing around 135,000 troops) with assistance from the UK (around 8,500 troops), while a number of other US allies such as Spain and the countries of central and eastern Europe committed limited force deployments. The differences between this predominantly unilateral military action and the multilateral humanitarian interventions in previous military engagements in Bosnia or Afghanistan appear stark. And yet, in the prosecution of intervention the transgression of Iraqi sovereignty could equally be marketed to sceptical domestic electorates as 'protective', 'humanitarian' and 'liberatory'. Despite the explicit geopolitical objective of the intervention, its justification was increasingly articulated through biopolitical vocabularies of protecting human life and improving the welfare of the Iraqi population.

Operation 'Iraqi Freedom' began on 20 March 2003. In an address to the American nation on the eve of the invasion, President Bush reiterated the enshrinement of state sovereignty and democratic legitimacy as a rationale for the use of military force: '[w]e have no ambition in Iraq except to remove a threat and restore control of that country to its own people'.[71] Vice President Dick Cheney agreed, suggesting 'we will, in fact, be greeted as liberators'.[72] This interpretation of the military action was reflected in the discursive division between state and society that threads through the Bush administration's reading of the intervention. Pathological interpretations of the war circulated through the Pentagon fantasy that the conflict would be one of 'decapitation', a surgical removal of Saddam Hussein's Ba'ath party from power leaving in place a liberated state architecture that could be remodelled into a democratic and neoliberal formation.[73] This objective is consistent with the logic of neoliberal intervention sketched in the previous section. The concept of 'decapitation' suggests a belief that Iraq suffered 'managerial' problems that could be rectified through the clear panorama afforded by US technical and military superiority coupled with the liberated energies of the Iraqi population.

As has become apparent, such 'clinical' predictions of regime change were naively optimistic. As a consequence of the invasion itself, low levels of US troop commitment, de-Baathification and the dismissal of the Iraqi army, the Iraqi state ceased to exist in any meaningful form.[74] Pieterse highlights the unprecedented shutdown of Iraqi state capabilities in the wake of the US-led invasion, '[the] entire government and civil service, armed forces, fire fighters, hospital staff, teachers

---

71   G. Bush, *President Bush Addresses the Nation* (19 March 2003) at <http://www.whitehouse.gov/news/releases/2003/03/20030319-17.html>.

72   MSNBC, *Meet the Press* (15 March 2003) at <http://www.msnbc.msn.com/id/3080244/>.

73   See BBC News, *US Strikes Target Saddam* (20 March 2003) at <http://news.bbc.co.uk/1/hi/world/middle_east/2867593.stm>.

74   T. Dodge, 'Iraqi Transitions: From Regime Change to State Collapse', *Third World Quarterly* 26/4–5 (2005) 705–21, p. 710.

and faculty were sent home and all production facilities stopped'.[75] In its enfeebled post-invasion form, the Iraqi state was not so much a 'rogue' as a 'collapsed' state. In April 2003 the US established the Coalition Provisional Authority (CPA) as an interim administration, initially headed by retired US Army general Jay Garner but replaced in May 2003 by career diplomat Paul Bremer. In addition to building security and 'essential services', the CPA sought reform of Iraqi governance through 'the establishment of an effective representative government, ultimately sustained through democratic elections'.[76] This objective included the drawing up of an interim constitution, creating open and transparent political systems and stimulating a vibrant civil society. But one of the most striking aspects of the CPA reforms was the dramatic neoliberalization of the Iraqi state. In September 2003, Bremer passed the infamous Order 39 that decreed that 200 Iraqi companies would be privatized; foreign firms would be able to retain 100 percent ownership of Iraqi industries and these same firms would be allowed to repatriate 100 percent of profits.[77]

The only measurable target the CPA set was the 'restoration of full sovereignty to the Iraqi people', a goal the CPA declared it had achieved at the point when it ceased to exist. Though it seems unclear how it could fail in such a limited objective, there was much celebration from the Bush administration when 'sovereignty' was handed to the US appointed Iraqi Governing Council (IGC) on 28 June 2004. This moment led President Bush to remark, 'after decades of brutal rule by a terror regime the Iraqi people have their country back'.[78] This stage-managed transfer of sovereignty encapsulates the geopolitical narratives examined in this chapter in three ways. Firstly, the 'return to the people' refrain suggests that US military intervention is innately democratic, that despite the removal of Iraqi industries out of public and into private hands the intervention can still be positioned as handing 'their country back'. Secondly, the sovereignty handover relies on an assumption of Iraq's territorial integrity despite increasing sectarian violence and incursions by the military forces of neighbouring countries such as Turkey and Iran. This unquestioning reliance on the territory of Iraq demonstrates the continued reproduction of the sanctity of state borders as containers of power within US security discourses (see Chapter 2), despite new economic measures granting the repatriation of company profit and legislating for foreign ownership of Iraqi firms. Thus the image of individual state sovereignty conflicts with the

---

75   J. Pieterse, 'Neoliberal Empire', *Theory, Culture and Society* 21/3 (2004) 119–40, p. 130.

76   Coalition Provisional Authority, *Historic Review of the Coalition Provisional Authority* (Baghdad: Coalition Provisional Authority 2004) p. 2.

77   See Harvey, *Brief History of Neoliberalism* (note 28); N. Klein, 'Baghdad Year Zero: Pillaging Iraq in Pursuit of a Neocon Utopia', *Harper's Magazine* (September 2004) pp. 43–53.

78   G. Bush, *President Bush Discusses Early Transfer of Iraqi Sovereignty* (28 June 2004) at <http://www.whitehouse.gov/news/releases/2004/06/20040628-9.html>.

reality of transnational interconnections set within unequal relations of power, some legitimate, others not. Finally, the idea of 'handing' a country back reinforces the notion that the US occupies a privileged moral position in identifying and correcting failed state sovereignty. In both planning, execution and post-conflict reconstruction, the US security strategies were structured around a supposed privileged geopolitical vista, the imagined ability to interpret the world with clarity unavailable to others.

The examples of Afghanistan and Iraq demonstrate the importance of situating declarations of 'state failure' beyond the abstracted threat cited in the national security strategies. As I have argued, declarations of state failure should be understood as productive speech acts structured around the primacy (but also indeterminacy) of statehood to the geopolitical imagination of current US security strategies. Rather than neutral reflections of an objective reality, such declarations of state failure serve to incite interventions on the basis of imagined hierarchies of competence and compliance. This reterritorialization of threat absolves the adjudicators of statehood from complicity in immoral or illegal activity by confining deviance to the borders of the problematic state itself.

Conclusion

> There was a time when two oceans seemed to provide protection from problems
> in other lands, leaving America to lead by example alone. That time has long
> since passed. America cannot know peace, security, and prosperity by retreating
> from the world. America must lead by deed as well as by example.[79]

The official security strategies of the US and UK are replete with geopolitical templates of connection and disconnection mapped onto differing interpretations of moral obligation. In this chapter I have examined one such spatial imaginary: the problematic state, whether fragile, failed or rogue. While these concepts emerged out of attempts to make sense of the 'new security environment' of the 1990s, they have undergone further iterations in the context of the war on terror. This new geopolitical imaginary invokes both the supposedly new 'networked' and 'deterritorialized' threat and the enduring statecentric character of power in the contemporary world.

Rather than portraying state failure as a detached assessment of the incapacity of individual states, I have offered a critical analysis that draws attention to the situated nature of adjudications of state sovereignty and their grounding in neoliberal and neoconservative understandings of state functions. Contributing to a growing body of work in critical geopolitics, this argument highlights the practices and interventions *enabled* by a particular writing of global space.[80]

---

79   The White House (2006) (note 26) p. 47.
80   For example Kuus (note 16).

In the case of both Afghanistan and Iraq, adjudications of statehood served as flexible placeholders onto which political and moral justifications for intervention could be overlaid. As I have argued, while these justifications may have enrolled biopolitical concerns of humanitarian plight, they were (and are) also mapped onto geopolitical narratives of ineffective or threatening state sovereignty. The fragile, failed and rogue state labels, then, serve to mask political practice behind technical and rhetorical renderings of state competence, severed from geographical and historical networks within which any given territory is entrenched. In doing so, the illusion of pluralistic political order structured around horizontally arranged sovereign states is reproduced.

If we disentangle the biopolitical and geopolitical dimensions of adjudications of statehood we are better placed to arbitrate on the reasons why discourses of state failure endure. The examples discussed in this chapter suggest that in geopolitical terms declarations of state failure are attractive to powerful actors in view of the ways in which they territorialize threat, deviance and difference. Taken at an international scale, the imagery of state failure geographically limits the causes and effects of the collapse of state institutions and reproduces the territorial self-esteem of hegemonic state actors. While interventions are routinely justified in biopolitical terms, the ability of the interventions examined in this chapter to improve the material welfare of targeted populations is less apparent. The example of Iraq demonstrates that despite the discourse of humanitarian protection, the post-intervention solutions have failed to stem human suffering, displacement, injury and death. Where the interventions were justified to domestic electorates in biopolitical (protective, humanitarian) terms, the subsequent experience questions the utility of military intervention to fulfil such objectives. In short, vocabularies of state failure persist in the face of failures to enact corrective policies since they perform a valuable geopolitical role within the development/security discourses of powerful state actors.

Finally, the examples of Afghanistan and Iraq demonstrate that despite the different nature of the security threats they were held to pose, the post-intervention solutions converge on the need to create neoliberal, neoconservative states. Policy makers rebut arguments of 'imperialism' and 'American Empire' through claims that intervention marks a temporary and necessary step, a bitter pill of a seemingly similar formulation as the 'austerity measures' and 'shock therapy' of IMF structural adjustment.[81] Where scholars have identified the subsequent states as neither democratic nor sovereign, such virtuous ideals seem misplaced.[82] The democratic rhetoric projects the failure in Iraq onto the individual competences of the Iraqi population and the government they have elected, the concept that

---

81   For example see M. Ignatieff, *Empire Lite: Nation-Building in Bosnia, Kosovo and Afghanistan* (London: Vintage 2003).

82   For a discussion of democratization in a Bosnian context see A. Jeffrey, 'The Geopolitical Framing of Localized Struggles: NGOs in Bosnia and Herzegovina', *Development and Change* 38/2 (2007) 251–74.

through the establishment of electoral competition the future of the state is 'in their hands'. This presentation of post-intervention states as containers of fate reflects the redeployment of statist language where failure is again territorialized to distance and absolve intervening agencies from their own complicity and responsibility.

Chapter 4
Colonizing Commemoration: Sacred Space and the War on Terror[1]

Nick Megoran

Introduction

On 11 September 2002 US President George W. Bush addressed mourners who had assembled outside the Pentagon building, just outside Washington DC, for a service of commemoration. Together they remembered the men, women and children killed when al-Qaeda operatives careered a hijacked plane into it exactly a year earlier. They also celebrated the building's restoration, and re-dedicated its use.

The commemoration began with a prayer by Major General Gunhus, Chief of Chaplains, US Army, who besought the (non-Trinitarian) Almighty to 'Fortify our inner courage as we hold the moral high ground', and re-dedicated the restored Pentagon to Him with the words, 'Now, Lord, we offer this sacred ground to you, and ask your blessing upon it.'[2] Following patriotic solo songs, and speeches by General Myers, Chief of Staff, and Donald Rumsfeld, Defence Secretary, President Bush himself took the podium. Reprising his themes of a struggle between people who love dignity and peace against those who murder without conscience, he assured the grieving military personnel present that 'Wherever [you are] sent in the world, you bring hope, and justice, and promise of a better day' as the US fought for 'peace in the world'. He then spoke of 'terrorists and dictators who plot against our lives and liberty', implicitly referencing the bogus claims of a link between Saddam Hussein and Osama bin Laden that his administration had used

---

1   Some material from this chapter is drawn from an article originally published by the author in 2006, 'God on our side? The Church of England and the geopolitics of mourning 9/11', *Geopolitics* 11/4 (2006) 561–79, and is reprinted by the permission of the publisher, Taylor and Francis Ltd. I acknowledge with much gratitude the kind assistance of the Very Revd Dr John Moses, formerly Dean of St Paul's Cathedral; Mr Dan Sreebny and Mr Mort Dworken, formerly of the Embassy of the USA in London; Sir Malcolm Ross and Mr Stuart Neville, the Lord Chamberlain's office, Buckingham Palace; and Lord Carey, formerly Archbishop of Canterbury. I would also like to thank the editors, Klaus Dodds and Alan Ingram, for their encouragement to write this chapter and their extremely insightful comments on earlier drafts.

2   Broadcast on BBC 1 (11 September 2002).

to justify the impending invasion and occupation of Iraq. Urging Americans not to forget 9/11, Bush declared, as the Pentagon building was re-dedicated, that 'we renew our commitment to win the war that began here'. Thousands of American and British service personnel were subsequently dispatched to two war theatres. Two years later, in May 2004, as that 'war' had been extended to Iraq, British war graves were vandalized in France, Iraq and Gaza, as a protest against the alleged torture of Iraqis by British soldiers, who were enforcing an occupation of Iraq of which George W. Bush was the chief architect.[3]

These different incidents illustrate how the war on terror and resistance to it manifest themselves in a variety of geographical settings by creating and colonizing sites of memory. These may be tangible sites, whose materiality is intentionally preserved over time, such as overseas British war cemeteries maintained by the Commonwealth War Graves Commission. They may also be temporary and intangible, such as those created outside the Pentagon on 11 September 2002. In both cases they may be sacred spaces, invoking through symbols, inscriptions or performances the watchfulness of a divinity.

This chapter explores the ways in which the nascent war on terror colonized the sacred space of St Paul's Cathedral, London, for a 'Service of Remembrance with the American Community in the UK', on 14 September 2001. It asks a simple question: in what way was this event inherently geopolitical? Following work by historians of war memorialization, it contends that public mourning is ambiguous – expressing genuine grief, whilst also being inherently political. Following Judith Butler, it suggests that the ability to publicly mourn the deaths of certain people dehumanizes others, thus perpetuating cycles of violence, by inserting the USA at the top of a 'hierarchy of grief'. On this basis, this chapter argues that, although most of the parties involved in organizing the service sincerely believed that they had crafted an apolitical event to enable grieving and provide comfort, the service articulated a geopolitical narrative. Crucially, it contends that, in a moment of indeterminacy when explanations of the events of 9/11 were contested in society at large, the service resonated with politically conservative voices in the US and UK that immediately called for a military response.

As an investigation, the chapter is based upon a critical geopolitical discursive reading of the service, as broadcast on British television and upon interviews with key participants in that service and those involved with its organization. Following a theoretical discussion of the nature and political significance of public mourning and commemoration as well as an outline of the service, there are two main empirical sections. The first identifies the multiple ways in which the service can be considered as geopolitical and the significance of St Paul's Cathedral as a site for the production and inscription of geopolitical knowledge. The second explores

---

3   'Iraqis Desecrate British War Cemetery', *The Daily Telegraph* (20 May 2003) at <http://www.telegraph.co.uk>; 'British War Graves Vandalised', *BBC News* (29 July 2003) at <http://www.bbc.co.uk>; 'Gaza Commonwealth Graves Vandalised in Protest', *The Namibian* (11 May 2004) at <http://www.namibian.com.na>.

how the service was constructed and my attempts to reconcile a geopolitical reading of the event with the organizers' explicit denial of geo-political content.

Geopolitics, Religion, Commemoration

Geography is about what Bonnett identifies as *Homo sapiens'* 'consistent desire to order their world, to find *meaning* in it'.[4] Geopolitics, the theoretical tradition within which this essay is located, is about the ordering and control of global space: it sets up places and regions in an imaginative geography, designating them as entities and imbuing them with qualities, providing a discursive framework within which wars can be thought and fought.[5]

The political geographer Simon Dalby applies such a (critical) reading of geopolitics to the immediate aftermath of the 9/11 attacks. In these uncertain times, he argues, 'the political ambiguities left a discursive and political space open for political leaders to fill with their specifications of events and appropriate responses to this new geopolitical situation'.[6] The events could have been specified as a crime, necessitating careful international collaborative police and security service investigation, or as the unintended consequence of earlier imperial entanglements. Indeed, whilst different explanations were offered for the events at the time,[7] critical commentators 'were effectively silenced by those calling for bipartisan and unqualified support by US leaders for the 'war on terrorism',[8] and a range of alternative options for responding to the events were sidelined and subsequently forgotten.[9] By designating the US as an innocent victim of unprecedented, unprovoked, motiveless evil, little apparent room was left for a discussion of the reasons that might have caused the violence. This de-legitimatization of alternatives mattered profoundly, argues Dalby, because these geopolitical specifications helped determine whether bombs would be dropped, and on whom.[10]

Locating itself within this theoretical tradition, this essay intersects with two topics rarely engaged by critical geopolitics. The first is religion. Although Thomas is surely correct in identifying that 'the global resurgence of religion'

---

4   A. Bonnett, *What is Geography?* (London: Sage 2007) p. 7.

5   P. Reuber, 'The Tale of the Just War: A Post-structuralist Objection', *The Arab World Geographer* 6/1 (2003) 44–6.

6   S. Dalby, 'Calling 911: Geopolitics, Security and America's New War', *Geopolitics* 8/3 (2003) 61–86, p. 62.

7   J. Taylor and C. Jasparo, 'Editorials and Geopolitical Explanations for 11 September', *Geopolitics* 8/3 (2003) 217–52.

8   S. Brunn, '11 September and its Aftermath: Introduction', *Geopolitics* 8/3 (2003) 1–15, p. 5.

9   C. Dahlman and S. Brunn, 'Reading Geopolitics Beyond the State: Organisational Discourse in Response to 11 September', *Geopolitics* 8/3 (2003) 253–80.

10   Dalby (note 6).

has had profound implications for the study of international relations,[11] political geographers have long been reluctant to engage with the topic. This situation is belatedly being rectified by a trickle of recent publications,[12] but these are largely focussed on either Islam or on Christianity in the USA. As I seek to demonstrate here, in the context of national services of commemoration, religion is also geopolitically significant in the UK.

The second topic is the commemoration of war dead. Historians of public mourning for wartime deaths have argued that it is highly ambiguous, providing both an outlet for grief and a means of coming to terms with tragedies, as well as always being political.[13] It is political in that it demonstrates whose loss a society considers worth marking, offers interpretations and explanations for the deaths of those being mourned, and engenders debate and dispute about how deaths should be commemorated. As Adrian Gregory has shown in his study of debates about the aftermath of the First World War in Britain, formal commemoration of the dead was sought by grieving widows and mothers, the military, and anti-war groups for very different reasons.[14] Jay Winter considers that the reason for the popularity of Sir Edwin Lutyens' Cenotaph memorial in London was its lack of religious or national symbolism and its absence of celebration, enabling it to become the focus of a range of different private thoughts.[15] Geographers have insisted that because memorials inscribe space with particular meanings, for example negotiating the divisions of a nation in mourning[16] or demonstrating national values in a tangible imprint on the landscape,[17] commemoration is inherently geographical. However, the work of both historians and geographers is largely confined to studies of material

---

11   S. Thomas, *The Global Resurgence of Religion and the Transformation of International Relations: The Struggle for the Soul of the Twenty-First Century* (Basingstoke: Palgrave Macmillan 2005).

12   J. Agnew, 'Religion and Geopolitics', special issue of *Geopolitics* 11/2 (2006) 183–91; Jason Dittmer, 'Religious Geopolitics', *Political Geography* 26/7 (2007) 737–39.

13   Key texts include: D. Cannadine, 'War and Death, Grief and Mourning in Modern Britain', in J. Whaley (ed.), *Mirrors of Mortality: Studies in the Social History of Death* (London: Europa 1981); J. Winter and E. Sivan (eds), *War and Remembrance in the Twentieth Century* (Cambridge: Cambridge University Press 1999).

14   A. Gregory, *The Silence of Memory: Armistice Day 1919–1946* (Oxford: Berg 1994).

15   J. Winter, *Sites of Memory, Sites of Mourning: The Great War in European Cultural History* (Cambridge: Cambridge University Press 1995) pp. 103–4.

16   N. Johnson, 'The Renaissance of Nationalism', in R. Johnston, P. Taylor and M. Watts (eds), *Geographies of Global Change: Remapping the World* (Oxford: Blackwell 2002) 130–42, pp. 138–9.

17   L. Kong, 'International Review: Cemeteries and Columbaria, Memorials and Museums: Narrative and Interpretation in the Study of Deathscapes in Geography', *Australian Geographical Studies* 37/1 (1999) 1–10, pp. 4–5; M. Morris, 'Gardens "For Ever England": Landscape, Identity and the First World War British Cemeteries on the Western Front', *Ecumene* 4/4 (1997) 410–34, p. 424.

war memorials. Although exceptions exist, transitory spaces of commemoration have been largely overlooked.[18]

Mourning 9/11

Social theorists and media scholars, amongst others, however, have not been slow to explore the significance of the ways in which the 9/11 attacks have been remembered and commemorated. A number of such analyses concur in the contention that the attacks exposed a vulnerability that could have opened a space for critical reflection on US foreign policy, but that moment was foreclosed by the nature of the remembering of the attacks. This is an argument that journalist Susan Faludi makes in a book about 'what 9/11 revealed about America'.[19] Jenny Edkins articulates the same point in an article that offers a four-fold classification of ways of remembering the trauma of seeing (in person or via television) the mass killing in the World Trade Centre. These began with *securitization*, the recollection of the trauma as an attack on America demanding a military response, and the *criminalization* of the attacks as the work of villains to be pursued through military-style police work. She claims these were the dominant two ways of remembering the attacks. More hopefully she viewed a third response as *aestheticization* of the trauma through services of remembrance and candlelit vigils, which might either lead to foreclosure towards securitization or keep the space of trauma open for more reflectively self-critical responses. Finally, she identified *politicization*, the recollection of the attacks not as a general assault on America but as a specific trauma, enabling critical reflection on US foreign policy. Edkins claims that whilst the US government sought to memorialize the attacks for securitization (much as President Bush did during the 2002 commemoration of the Pentagon attack, cited in the opening vignette), many families of victims preferred to politicize them and baulked at the marshalling of their grief in support of the war on terror.[20]

This politicization of the memory of loss by grieving families is discussed by Maja Zehfuss in her jarring call to 'forget September 11th': a course of action she considers better than allowing its memory to be used to justify militarism. She adduces the example of the relatives of Craig Scott Amundson, who died in the Pentagon. His widow, Amber Amundson, movingly addressed President Bush with the words:

---

18   M. Azaryahu, 'The Spontaneous Formation of Memorial Space: The Case of Kikar Rabin, Tel Aviv', *Area* 28/4 (1996) 501–13; K. Hartig and K. Dunn, 'Roadside Memorials: Interpreting New Deathscapes in Newcastle, New South Wales', *Australian Geographical Studies* 36/1 (1998) 5–20; B. Fletcher, 'Anglicanism and Nationalism in Australia, 1901–1962', *Journal of Religious History* 23/2 (1999) 215–33.

19   S. Faludi, *The Terror Dream: What 9/11 Revealed About America* (London: Atlantic Books 2007).

20   J. Edkins, 'Forget Trauma? Responses to September 11', *International Relations* 16/2 (2002) 243–56.

If you choose to respond to this incomprehensible brutality by perpetuating violence against other human beings, you may not do so in the name of justice for my husband. Your words and imminent acts of revenge only amplify our family's suffering, deny us the dignity of remembering our loved one in a way that would have made him proud and mock his vision of America as a peacemaker in the world community.[21]

On another occasion, she read out a letter in front of the White House, saying:

So, Mr President, when you say that vengeance is needed so that the victims of 9/11 do not die in vain, could you please exclude Craig Scott Amandson [sic] from your list of victims used to justify further attacks.[22]

In her book on the USA and the war on terror, *Precarious Life*, Judith Butler repositions the debate on remembering the dead of 9/11 around the questions, 'Who counts as human?', 'Whose lives count as lives?' and 'What *makes for a grievable life*?'[23] She affirms the political importance of grieving in furnishing a sense of political community by foregrounding relational ties that reveal fundamental dependencies and ethical responsibilities. She observes, however, that a 'hierarchy of grief' structures the ability to mourn and thus the inclusion or exclusion of individuals from the human category. The dead of 9/11 were named and grieved individually while 'There are no obituaries for the war casualties that the United States inflicts.'[24] She cites as an example a Palestinian US citizen who submitted obituaries to the *San Francisco Chronicle* of family members killed by (US-backed) Israeli forces, yet found the submissions rejected on the basis that they might cause 'offence'. The obituary, she concludes, is not just about nation building and the construction of a sense of 'we', but disqualifies other lives as being not worthy of being noted let alone commemorated.

This body of literature extends our understanding of what commemoration *does*: at best, as in Butler, it recognizes the importance of grieving whilst problematizing and politicizing its omissions. However, its key weakness is a lack of detailed, empirical exploration of just *how* sites of memory are created and by whom. Within this context, this chapter bridges gaps in these literatures by drawing the theoretical strands of geopolitics, commemoration, and the mourning of 9/11 together in an investigation of the service at St Paul's Cathedral service on 14 September 2001. Exploring the geopolitics of the service, the chapter argues that the actual service illustrates Butler's 'hierarchy of grief' in a form of remembering

---

21   Cited in M. Zehfuss, 'Forget September 11th', *Third World Quarterly* 24/3 (2003) 513–28, pp. 524–5.

22   Ibid.

23   J. Butler, *Precarious Life: The Powers of Mourning and Violence* (London: Verso 2004) p. 20.

24   Ibid., p. 34.

that leaned towards Edkins' conception of trauma *securitization* rather than its critical *politicization.*

The Service of Remembrance, 11–14 September 2001

The service occurred at an extraordinarily tense moment. Still only days after the airborne slaughter inflicted by 19 al-Qaeda operatives stunned television audiences around the world, little was known about the attackers and their motives. American and British explanations of the reasons for and correct responses to the attacks had not yet been codified as the war on terror, but were being widely debated and speculated about in the media. It was in this extremely tense context that British churches joined those in the US and elsewhere in holding services of prayer and commemoration on 14 September, linked to a 3-minute silence observed by an estimated 800 million people worldwide.[25] The major British event, at St Paul's cathedral, was the focal point of national mourning, immediately following a special parliamentary session in which the Prime Minister, Tony Blair, made a hawkish speech preparing Britain for war, insisting on the need for an 'implacable and long fight against terrorism', to save Britain from possible future nuclear strikes.[26] The *Times* noted that Mr Blair was 'more aggressive than other European leaders', observing that he left no doubt about Britain's readiness to join the 'war'.[27] Thus, it was already clear before the service began that the British government was positioning itself to join the USA in military retaliation for the attacks.

Although the congregation assembled to observe the 3-minute silence at 11am, the Service of Remembrance with the American Community in the UK began at 12pm on 14 September. By then, thousands of mourners had packed the streets from St Paul's down Ludgate Hill, an unusual occurrence that made a great impression on cathedral staff.[28] The congregation largely consisted of members of the public who had queued since 5.30am, but also the Queen with Princes Philip and Charles, senior members of the government and opposition, all living past prime ministers and representatives of London's major financial institutions and firms, including those who had lost staff in the Twin Towers.

The service proper began with the singing of the US National Anthem, and then proceeded with the lighting of a candle, a bidding prayer by the Dean, Dr John Moses, the Lord's Prayer, intercessory prayers, hymns, anthems, a psalm and scripture lessons read by the Duke of Edinburgh and the US Ambassador, William Farish. George Carey, Archbishop of Canterbury, gave a six-and-a-half minute address. The service concluded with the singing of *Glory, Glory, Hallelujah!*, a final blessing and one verse of the National Anthem. The service was broadcast

25 'A Silent World, United in Grief', *Independent* (15 September 2001) 1.

26 'America's Might Turns on Arab Terrorist Havens', *The Times* (15 November 2001) 1.

27 Ibid.

28 Interview, John Moses, St Paul's Cathedral (29 April 2003).

live, with running commentary provided by the veteran BBC television journalist, David Dimbelby.

The Church of England and the 14 September Service

Responses to 9/11 were not predetermined and this space of trauma enabled different interpretations of the attacks to be made. That the service had the potential to follow through Edkins' perspective on *aestheticization*, if not *politicization*, provides a reminder of the alternative uses of memory that existed alongside the rapidly emerging Bush-Blair *securitization* paradigm. This potential is clear from the ecclesiastical and theological context of the 14 September 2001 memorial service.

Ecclesiastically, while the Church of England has been shaped by its jingoistic support of the First World War, it has displayed a willingness to diverge from British national geopolitical narratives. In the early 1980s the Archbishop of Canterbury, Robert Runcie, angered Margaret Thatcher by refusing to provide an unequivocal celebration of the 1982 Falklands War. Instead he crafted a service in St Paul's that included prayers in both English and Spanish for grieving relatives in Britain and Argentina, emphasizing that Argentines 'are indeed like us'.[29] In 1983 the church adopted, to the chagrin of the Thatcher government, a position of unilateral nuclear disarmament in its controversial report, *The Church and the Bomb*. George Carey, Runcie's successor to the See of Canterbury, as first incumbent of the position in modern times to have working class origins and not to have been educated at Oxford or Cambridge, was not party to the traditional British establishment. He had also sought to establish links with Muslims in Britain and abroad and deservedly earned a reputation as a builder of bridges and the promoter of a deeper understanding of Islam.

Theologically, New Testament scholarship in the 1990s that engaged with post-colonial theory was increasingly recovering sensitivity to the imperial context of the birth of Christianity and the anti-militarism, or rather positive commitment to non-violent confrontations in the face of an oppressive colonial regime, that characterized the early church before its co-option by the Roman Empire in the fourth century.[30] This literature extrapolates what it postulates as the early Christian refusal to be co-opted by Roman imperialism to a call for the same sceptical stance to contemporary American empire[31] – or, as Jack Nelson-Pallmeyer strikingly puts it, to 'save Christianity from empire' by exposing how Christian support of US foreign policy has subverted and distorted the faith (see

---

29 A. Hastings, *Robert Runcie* (London: Mowbray 1991) pp. 89, 183–6.

30 A good example is the work of Richard Horsley. See for example *Jesus and the Spiral of Violence: Popular Jewish Resistance in Roman Palestine* (Cambridge: Harper and Row 1987).

31 R. Horsley, *Jesus and Empire: The Kingdom of God and the New World Disorder* (Minneapolis: Fortress Press 2003).

also Chapter 12).[32] Thus, for both ecclesiastical and theological reasons, it might have been anticipated that the service could have opened a space for reflection on alternative ways of remembering the events to those being developed by the US and UK governments.

The Geopolitics of the Service

Public mourning is ambiguous: as well as addressing shock and grief and the searching emotional and spiritual questions that arise at times of loss and tragedy, it is also political. The following sections will explore the geopolitical imagination enacted in the service.

Hierarchies of Grief

John Agnew has defined a core exercise of geopolitics as 'geographical framing', where the world is 'actively "spatialized", divided up, labelled, sorted out into a hierarchy of places of greater or lesser "importance"'.[33] The very fact of holding the service was an example of such geographical framing. The presence of so esteemed a congregation demonstrated that the events of Tuesday that week were of the utmost importance. It was widely reported that the usually reserved British monarch wept, the *Daily Mail* claiming it was the first time that she had cried in public since the decommissioning of the royal yacht *Britannia*.[34]

Writing to *The Independent*'s letters column the following day, Natasha Walter admitted, 'All these tears make me uneasy.' It is fine to mourn, she continued, 'But where have been our three-minute silences in recent years for the dead of Rwanda or Srebrenica or Sierra Leone?'[35] The point she raised is important. Casualty numbers are a crude way of marking significance, but the number of dead on 9/11 is dwarfed by the millions who were killed in conflict in the preceding years in Central Africa, Ethiopia/Eritrea, and numerous other places. The UN reckoned that 5,000 children were being killed in Iraq *each month* at that time as a result of sanctions enforced largely by the US and UK. But death on this scale is not confined to conflict. In January 2001, 20,000 Gujuratis perished in one of the deadliest earthquakes ever to have struck India,[36] whilst four months after 9/11, over one thousand Nigerians perished in the appalling Oke-Afa munitions blast

---

32   J. Nelson-Pallmeyer, *Saving Christianity from Empire* (London: Continuum 2005).

33   J. Agnew, *Geopolitics: Re-Visioning World Politics* (London: Routledge 2003) p. 3.

34   'She Sang Along with the American Anthem. And, Like Everyone, She Wept', *Daily Mail* (15 September 2001) 2–3.

35   Letters, *Independent* (15 September 2001) 5

36   D. Lewis (ed.), *The Annual Register of World Events, 2001* (Bethseda: Kessing 2002).

and the consequent stampede into the canal.[37] Although these Commonwealth states have closer formal historical-constitutional links with the UK than the USA, there were no similar public demonstrations of grief.

The attacks of 11 September had a colossal impact on Britain, bringing everyday life to a standstill. Radio 4 comedy programmes were ditched, sporting events cancelled, and top-selling tabloid The *Sun* even suspended its infamous 'Page 3' topless young woman slot. It was apparently deemed appropriate to continue ogling half-naked women and laughing when the other deaths occurred, but not when people perished in America on 9/11. In holding such a high-profile service for 9/11 and not for these other tragedies, the service reinforced the notion that the lives of Americans, and others living in America, matter more than lives elsewhere.[38] I have seen no evidence that, personally, clerics in the Church of England believed that (see discussion below based on interviews) but the notion that the suffering of America on 9/11 was uniquely terrible was important to the US and UK governments' case for military retaliation by invading Afghanistan and subsequently Iraq. St Paul's has not subsequently hosted national services of commemoration for those in Iraq and Afghanistan killed by US and UK forces. As Butler put it, those killed in the USA are more 'grievable' than those killed by the USA and this type of mourning perpetuates violence by presenting the US deaths as a trauma to be avenged against America's ungrievable enemies.[39]

Geopolitical Alliances

The service rehearsed a geopolitical vision of the identity of Britain as, alongside the US, a champion of liberty in a dangerous world. It posited a special link between the UK and the US that was emotional yet also political and military. John Moses's opening words were, 'We come together as members of the free world …'. This beginning of a service of Christian worship with an overt reprisal of a Cold War ideological trope encapsulated the geopolitical subtext of the whole event.

The order of service embodied this unique relationship. Norman Cooley, an American working for a law firm whose business had an office in the World Trade Centre, processed up the nave with a large cross. Lauren Willoughby, an American student, lit a candle of remembrance, and prayers were jointly said by Philip Butlin, Canon in Residence, and Mrs Marcia Molloy, an American lay reader based in London. The latter was a non-ordained member of the laity formally licensed by the Church of England to officiate and preach at certain services. The service itself

---

37   'Memorial Holds for Jan 27 Bomb Blast Victims at Oke-Afa', *Vanguard* (28 January 2003) at <http://www.vanguardngr.com>, accessed January 2003.

38   This may also be due to the apparently simple if important fact that far more people in the UK have been to the US and specifically New York and that political, economic and cultural connections are far stronger than other parts of the world such as Rwanda.

39   Butler (note 23) Chapter 2, 'Violence, Mourning, Politics'.

was sandwiched between singing of the American and British national anthems – the first time that the former was played at St Paul's. Thus, through a variety of devices, the service evocatively linked Britain and America together. As the *Church Times* rather touchingly put it, 'It looked like two families, each with its matriarch and patriarch: the Queen, in black and visibly moved, with the Duke of Edinburgh; and the US Ambassador, William Farish, with his wife Sarah. The patriarchs read the lessons ... '.[40]

Such a connection, embodied in those performing the service as much as in the words sung and spoken, was indeed entirely appropriate for the service, and a moving way to express sympathy. However, in the wider geopolitical context, this linkage was not only sentimental or emotional, but also military and political and, was to play a predictably important role in the war on terror.

Geopolitical Explanations for 9/11

The only part of the service that was suggestive of an explicit explanation of the Tuesday attacks was the Archbishop of Canterbury's sermon. Avoiding any suggestion that the attacks might have either historical or political contexts, Archbishop Carey asserted that the attackers were 'evil' – a word supplemented with adjectives such as 'despicable'. Lacking comprehensible goals or grievances, the attackers were apparently motivated only by wickedness itself, perpetrating 'a senseless evil', as he put it. However, the Archbishop did go on to identify a target – liberty. He proclaimed that the American people had suffered an 'assault on their freedom'.

Central to this explanation was a binary geographical imagination of two different realms. On the one hand, there was America and 'the free world'. America was portrayed as an innocent victim, being 'a noble community of values in which we are proud to share, values like tolerance and compassion, justice and mercy'. The Archbishop stressed that as 'a senseless evil had been perpetrated against America and against the free world', this was an attack upon all who shared American values.

The attackers, on the other hand, inhabited a realm of 'evil' and 'darkness'. At odds with all that the free world valued, they attacked liberty itself and their actions were 'barbaric'. The labelling of one's opponents as barbarians reflects a trope historically deployed by Europeans to insist upon their supremacy over non-Europeans,[41] particularly apparent in European discourse about the Middle

---

40   'St Paul's Hears Words of Hope', *Church Times* 7231 (21 September 2001) p. 3

41   G. Ó Tuathail, *Critical Geopolitics: The Politics of Writing Global* Space (London: Routledge 1996) p. 104.

East[42] and other majority Muslim areas.[43] The Archbishop of Canterbury was, whether wittingly or unwittingly, signalling an idea which has legitimized Western militarism in the Middle East – and would go on to do so in Bush and Blair's war on terror. This is ironic, given what was said above about his prior attempts to engage with Islam.

The sermon did not merely posit a binary geopolitical division of good versus evil reminiscent of the Cold War: it gestured towards identifying America as a gleaming example of divine virtue on earth. It did this by two moves. Firstly, the Archbishop stated that the values that America embodies are those values, 'at the heart of the Christian faith and also of other great faiths'. But, uniquely, America is a beacon for other nations to look to, symbolized by America's most famous monument:

> As the twin towers of the World Trade Centre disappeared amid the smoke and carnage, across a short stretch of water another, older, American icon was not submerged. The September morning sun continued to shine on the Statue of Liberty, a torch raised like a beacon, a beacon of hope, and to millions around the world, a symbol of all that is best about America.

That the Statue of Liberty is an emblematic figure of the US is undeniable. However, to conflate it with America as 'a beacon of hope' is contestable, overlooking as it does the ambiguous nature of the motivations for and impacts of US foreign policy around the world. To acknowledge that is far from claiming that the USA is peculiarly evil – it is not. However, to describe it in the near-messianic terms that the Archbishop used is in turn to make an extraordinary statement in theological terms, which anticipated the foreign policy discourse of the Bush presidency.[44] This impression was perhaps further emphasized by the Archbishop's use of Biblical prophecy, quoting from Isaiah chapter 61, to illustrate his hope that America might rise up as a stronger nation.

The Coming Judgement

The Archbishop's words served to entrench the idea that America was uniquely good and those who had had attacked it uniquely diabolical. For the Archbishop, this conflict was part of a struggle of light versus darkness: 'as we battle with evil',

---

42   E. Said, *Orientalism* (New York: Vintage 1979); E. Said, *Covering Islam: How the Media and Experts Determine How We See the Rest of the World* (London: Routledge & Kegan Paul 1981).

43   A. Bichel, *Contending Theories of Central Asia: The Virtual Realities of Realism, Critical IR and the Internet.* Unpublished PhD dissertation, Department of Political Science, University of Hawaii, p. 181.

44   P. Singer, *The President of Good and Evil: Taking George W. Bush Seriously* (London: Granta Publications 2004).

he said of America's expected response. Having repeated that 'liberty has always been at the heart of the American vision', he went on to state:

> … that liberty must be defended. It is the awesome responsibility of the leaders of America now to decide how to respond to this evil inflicted upon their people, this assault on their freedom and security. The leaders of America need our prayers. May God give them wisdom to use their great power in such ways that further evil aggression is indeed deterred.

This statement does not specify what kind of response the Archbishop meant but it does raise the possibility as to whether the speech was hinting at what was to come to be known as the Bush Doctrine – preventing future attacks through deterrence and pre-emption. However, with the media rife with speculation about which countries America would attack, it would likely lead its hearers to assume a military response. The Archbishop was not making a statement of unbridled bellicosity – he urged the US and Britain to be 'guided by higher goals than mere revenge' and to keep 'values like tolerance and compassion, justice and mercy' in front of us 'like a beacon.' His words would seem to imply that he was cautioning in favour of a measured, considered military response, within the parameters of the just war tradition, but it is certainly not the antimilitarist statement that theologians such as Lee Griffith[45] and Michael Northcott[46] desired. In that sense, it was again closer to the positions of Bush and Blair that a strong response was needed and justified, but not one in haste or for revenge.

When Archbishop George Carey concluded his sermon, the congregation arose and launched into a stirring rendition of *Glory, Glory, Hallelujah!* (also known as '*The Battle Hymn of the Republic*') which begins:

> Mine eyes have seen the glory of the coming of the Lord:
> He is trampling out the vintage where the grapes of wrath are stored;
> He hath loosed the fateful lightning of his terrible swift sword:
> His truth is marching on.
> *Glory, glory, hallelujah!* …

The theological assertion of this patriotic Civil War hymn is clear from a later verse: 'As he died to make men holy, let us live to make men free / While God is marching on' is a Christological reference, imbuing the militant American nation with a soteriological mission in history. Clifford Longley identifies it as, 'plainly a battle hymn for an elect nation, a Chosen People'.[47] As Brian Walsh and Sylvia

---

45   L. Griffith, *The War on Terrorism and the Terror of God* (Cambridge: Eerdmans 2002).

46   M. Northcott, *An Angel Directs the Storm: Apocalyptic Religion and American Empire* (London: I.B. Tauris 2004).

47   C. Longley, *Chosen People: The Big Idea that Shapes England and America* (London: Hodder and Stoughton 2002) p. 329.

Keesmat argue, songs such as the Battle Hymn of the Republic 'provide a moment of ritual that gives religious legitimation to the American empire'.[48]

Place and the Embodied Inscription of Geopolitics

In his study of ceremony on Australian Remembrance Day, Stephen Muecke argues that its role in re-creating the nation, re-enacting a myth of origin through the music and ceremony in 'bodies full of affect' involves more than just conveying cerebral ideas: it is about the power of a vital force leading to the 'transformation of the bodies of those present at this ritual'.[49] It would be a mistake to exhaust the significance of the service with a textual reading of its geopolitical codes. The unique surroundings of St Paul's were no mere empty container for a public/ political event, but rather, in the impact of the service on sensuous human bodies, a vital element in the service and in informing interpretations and explanations of 9/11.

St Paul's regards itself as the unofficial national cathedral of Britain. As Stephen Daniels says of it, 'Launched by the unified efforts of Crown, Church, City and Parliament, the Cathedral was, from the beginning, seen as a symbol of national as well as civic renewal and ascendancy'.[50] To enter Sir Christopher Wren's great temple is to open one's aural, visual, and spatial senses to an extraordinary aesthetic experience. The sheer performance of a solemn religious service in this august setting is dramatic and may, even for those who are not Anglican, be moving and awe-ful. It lends an event occurring in the Cathedral a unique sense of significance. This feeling of occasion was compounded not only by the density of bodies packed within and around the cathedral, but that their number included 'Very Important Persons' at the apex of state, church, and commerce, literally embodying the establishment of Great Britain. The orchestration of these bodies through positions of processing, sitting, kneeling, and standing, to the accompaniment of sudden movements between silent reflection, prayer and dramatic reading, meditative music and thunderous proclamation of stirring hymns, evokes emotional and affective responses in participants (and also possibly television viewers). The Queen's rare public presentation of emotion is ample evidence of this.

The ambiguity between grief and politics at the heart of the service was illustrated by the music chosen: anthems and hymns that expressed grief and sorrow on the one hand and national and militaristic songs that contributed to the construction of a geopolitical imagination on the other. The songs sung during the service are of interest not simply because of their lyrics: they exercise an extra-

---

48   B. Walsh and S. Keesmaat, *Colossians Remixed: Subverting the Empire* (Milton Keynes: Paternoster Press 2004) p. 183.

49   S. Muecke, 'Travelling the Subterranean River of Blood: Philosophy and Magic in Cultural Studies' *Cultural Studies* 13/1 (1999) 1–17, pp. 2–3.

50   Stephen Daniels, *Fields of Vision: Landscape Imagery and National Identity in England and the United States* (Cambridge: Polity Press 1993), pp. 18–19.

lyrical power of evocation that is crucially dependent upon the environment in which they are played. Politically, Sabrina Ramet argues, music,

> brings people together and evokes for them collective emotional experience to which common meanings are assigned ... It gives them common reference points, common idols, and often a common sense of "the enemy".[51]

George Revill argues that music should not be analysed merely as text, but that its power in moving embodied subjects should be acknowledged.[52] Bob Woodward reported that television footage of the Coldstream Guards playing the Star Spangled Banner outside Buckingham Palace on 13 September made President Bush's National Security Adviser, Condoleezza Rice, weep.[53] Ms Rice is a Christian, but even President Bush's speechwriter, David Frum, a Jew, listening to a military band play the Battle Hymn of the Republic at the 14 September service in Washington National Cathedral, found himself stirred by what he described as 'the full throated anthem of Protestant righteousness militant'.[54] Revill's argument in general is illustrated by some of the coverage of the service in the press. The left-leaning *Guardian* newspaper wrote of the service:

> It was the playing of the American national anthem that did it. As the 19th-century song with its stirring words about the broad stripes and bright starts "giving proof through the night that our flag was still there" swelled through St Paul's Cathedral, a convulsive sigh echoed through the congregation.[55]

It was the stirring power of music on 'bodies full of affect' in this unique gathering in this unique place that produced such an effect. This context afforded the words of Archbishop Carey, delivered not as political opinion but with divine invocation, an authority and impact that merely issuing a statement or holding a press conference could not possibly have done. In order to understand the war on terror, it is necessary to take seriously the role of ritual performances in sacred

---

51   S. Ramet, 'Rock: The Music of Revolution (and Political Conformity)', in S. Ramet (ed.), *Rocking the State: Rock Music and Politics in Eastern Europe and Russia* (Oxford: Westview Press 1994) p. 1.

52   G. Revill, 'Music and the Politics of Sound: Nationalism, Citizenship and Auditory Space', *Environment and Planning D: Society and Space* 18/5 (2000) 597–13.

53   B. Woodward, *Bush at War* (London: Simon and Schuster 2002) pp. 63–4.

54   D. Frum, *The Right Man: An Inside Account of the Surprise Presidency of George W. Bush* (London: Wiednfeld and Nicolson 2003) p. 138.

55   'Tears and Sympathy as Millions Across Britain Stand Silent', *The Guardian* (15 September 2001) 3. Likewise, *The Independent* reported that the singing of a foreign anthem in 'Britain's national house of worship' was 'the most resonant expression of grief shared between two countries that it was possible to imagine', which showed that the 'special relationship' was not a tired cliché, but a reality ('Britain Stands Shoulder to Shoulder to Mourn US dead', *The Independent* [15 September 2001] 4).

space – or more accurately in specific sacred places – in re-inscribing geopolitical imaginations.

Responses to the Service

The geopolitical interpretation of the service offered in this chapter was precisely how the service was understood in the media. For the *Mirror* it, 'was not just a service of remembrance. It was a celebration of unity, delivering a message of defiance'.[56] On the morning of the service, the *Sun* announced the 3-minute silence on its front cover and urged support under the headline: 'God bless America: 3-minute silence today at 11am.'[57] An editorial made it clear what it thought the purpose of the silence and commemoration service was: a space to gather our thoughts, focus on the task, and prepare for 'score settling time'.[58] The *Daily Mail* observed that, 'More than a memorial', it was a statement of alliance between two peoples,[59] a military alliance that should use all means to 'crush the terror'.[60] The service and the words of the Archbishop of Canterbury were reported around the world, particularly in the USA.[61] Reception of the service in the UK and US media illustrates the contention of this paper: that the service enacted an ambiguity of meaning, both as an expression of grief and a search for sense and comfort, but can also be interpreted as a geopolitical text explaining why 9/11 occurred and its unique significance, preparing the discursive terrain for war.

The Arrangement and Organization of the Service

This chapter is not suggesting a dark conspiracy or malignant plan to script a service to aid the US and UK case for war. Both Mike Heffernan[62] and Mandy Morris[63] argue that it is not enough simply to explore the meanings associated with spaces of remembrance, but also to attempt to piece together some of the processes of their construction. Likewise, Jeremy Crampton and Gearóid Ó Tuathail insist

---

56   'United in Our Tears', *Daily Mirror* (15 September 2001) pp. 14–15.
57   'God Bless America: 3-minute Silence Today at 11am', *The Sun* (14 September 2001) p. 1.
58   'A Short Silence ... Then We Have a Job To Do', *The Sun* (14 September 2001) p. 10.
59   'She Sang Along', *Daily Mail* (note 34) pp. 2–3.
60   'Comment', *Daily Mail* (15 September 2001) p. 18.
61   For example, 'The UK', *Irish Times* (14 and 15 September 2001) 2; 'Global Grieving: World Unites in Time of Sorrow', *Washington Post* (15 September 2001) A19; 'Prayer: A Time to Mourn Across the Globe', *Los Angeles Times* (15 September 2001) A7.
62   M. Heffernan, 'For Ever England: The Western Front and the Politics of Remembrance in Britain', *Ecumene* 2/3 (1995) 293–323.
63   Morris (note 17).

that research in critical geopolitics that considers discourse alone is too narrow.[64] It is too easy to impute an inaccurate intention to the producers of texts and the choreographers of space.

In this section, therefore, I will consider *how* the service was arranged. For this, I interviewed representatives of core bodies involved in organizing and participating in the service. These were: the Very Revd Dr John Moses, Dean of St Paul's Cathedral; Mr Dan Sreebny, Minister Counselor for Public Affairs at the US Embassy in London; Mr Mort Dworken, Minister Counselor for Political Affairs, also at the US Embassy in London; from Buckingham Palace Sir Malcolm Ross, Controller of the Lord Chamberlain's office and his colleague Stuart Neville; and Lord Carey. Unfortunately, the British government, which was also involved in discussions over the service but did not ultimately play a role in organizing its details, did not respond to my requests for an interview with the relevant point of contact.

From these interviews, I was able to build up a comprehensive picture of how the service was arranged and why it took its eventual form. It would appear that, following an approach by the American Embassy to St Paul's early on Wednesday morning and in discussion with Downing Street and other actors, it was quickly agreed that a service of Remembrance should be held at St Paul's on the Friday, the day that similar events would happen across Europe and in America itself. John Moses told me that, 'It was essentially organized in three hours', by midday Wednesday.

Respondents indicated a remarkable degree of unanimity about the purpose and politics of the service. There was a shared understanding of its goal as being threefold: to grieve at a time of shock, express solidarity with the USA and to offer hope in the midst of despair. Reflecting on the role of St Paul's, as he saw it, to offer help to the grieving nation, John Moses said of the service that 'The primary purpose is providing space – architectural and liturgical space – so that people can come and grieve.' Just as all parties were agreed on the goals of the service, they were equally emphatic that the service was apolitical. As Malcolm Ross put it simply, 'We don't discuss politics'. Mort Dworken had been anxious in arranging the service that it would 'reach across the political spectrum', and in particular that Muslim representatives be included. Citing the example of 'Falklands triumphalism', John Moses was fully aware that in such services, 'the temptation is always to collude' and was proud of the Cathedral's success in maintaining, as he saw it, the political independence of the service.

This desire to focus upon the three goals identified and steer clear of politics can be demonstrated by two examples. George Carey, saying that the sermon, 'was not intended to be political', recounted that his colleagues persuaded him to remove some comments from a draft of his sermon about the need for Christianity and

---

64   A. Crampton and G. Ó Tuathail, 'Intellectuals, Institutions and Ideology: The Case of Robert Strausz-Hupé and "American Geopolitics"', *Political Geography* 15/6–7 (1996) 533–555, p. 553.

Islam to stand together against Islamic fundamentalism, as they considered it, 'too politicized'. Likewise, the US embassy questioned the inclusion of the *Battle Hymn of the Republic* in the order of service, as they thought that it was inappropriate for a commemoration service, the lyrics being, 'somewhat aggressive, militaristic' (Dan Sreebny). John Moses, however, explained to me that it was chosen because it was easily identifiable as an American national song and would be known by people in Britain. Its inclusion was thus not a sinister conspiracy between Bush and Blair.

Two academic writers, Katz and Liebes have argued that 'establishment events' are usually co-produced by broadcasters and organizers.[65] However, it is important to disaggregate the components of 'the establishment' if conspiratorial reasoning is to be avoided. Actors such as the Church of England, the US Embassy and Buckingham Palace, traditionally anxious to remain outside domestic political controversy, and particularly the owners of the event, St Paul's Cathedral, strove to ensure that the service would precisely *not* be interpreted as 'political'. There was no elite conspiracy. However, as I have argued throughout, the service not only could be interpreted as reprising a number of geopolitical tropes that harmonized with the emerging interpretation of George W. Bush, but that this was indeed how it was understood, at least by the popular media.

This apparent contradiction emerges from the ambiguities of a public service of commemoration that addresses grief and loss yet is also political, the working of taken-for-granted assumptions about how the world works that critical geopolitics is so adept at uncovering and vitally, different concepts of the 'political'. If 'politics' is seen as the formal issuing of explicitly partisan statements advocating certain courses of action, then the service was indeed apolitical. However, if 'politics' is considered as including the discourses that pervade social interaction and have meaning in precise contexts because of their *intertextuality*, or how they are interpreted in relation to other discourses in other contexts, the service did indeed script a geo-politics of the events.

It is impossible to claim that the service had a quantifiable impact on policy or public sentiment. Nevertheless, at a crucial moment, certain aspects of it dovetailed with discourses elsewhere in the public sphere to augment what was emerging in policy circles as a dominant understanding of the events of that terrible Tuesday morning. This geopolitical imagination of an innocent America attacked by depoliticized 'evil' helped make the projection of US and UK violence in Central Asia and the Middle East easier both for governments to legitimize and vindicate, and populations to countenance.[66]

---

65   E. Katz and T. Liebes, '"No More Peace!": How Disaster, Terror and War Have Upstaged Media Events', *International Journal of Communication* 1/1 (2007) 157–166, p. 164.

66   K. Dodds, *Global Geopolitics: A Critical Introduction* (London: Pearson Education 2005) p. 226.

Conclusion

The St Paul's service of Remembrance on 14 September 2001 was both an expression of grief and a geo-political intervention. It was clear by that stage that the UK government was positioning itself to engage with the US in a military response, premised upon the notion that the attacks expressed not political grievance but an irrational hatred of democracy and could only therefore be countered with violence. However, this interpretation was far from uncontested in Britain, the USA and the world at large. Occurring thus at a crucial moment, the service can be read as containing a geopolitical text that more closely approximated those voices advocating violence. It did this by employing a number of facets of geopolitics well documented by geographers, from the insertion of the USA into a 'hierarchy of grief' by the ranking of places as more or less important, to the reprising of national myths of geopolitical alliance. A particular geopolitics colonized St Paul's as a space of commemoration, despite the efforts of the organizers to script what they considered to be an uncontroversial, apolitical service.

A ritualized performance of mourning in national sacred space was part of a process that encoded both the 9/11 attacks and the subsequent US-led wars as 'the war on terror'. This (contingent) outcome was enacted not only at St Paul's, but at a series of commemorative performances invoking the divine presence at sites such as the US Congress, the rubble of the World Trade Centre,[67] and the Pentagon. Leading US evangelical, Jim Wallis, wrote that 9/11 could have been a 'teachable moment' for the USA when the remembrance of trauma occasioned an opportunity, a space, in which US foreign policy priorities and tactics could be critically considered and re-oriented towards a more peaceable engagement with the world.[68] On 14 September 2001, the Church of England missed the opportunity to create such a space. As the most public moment of British commemoration in the immediate aftermath of the attacks, that was not without consequences. The outcome of the geopolitical designations adopted and the subsequent courses of actions chosen by the US and UK governments included the disastrous and immoral invasions of Afghanistan and Iraq as part of a process of remilitarization long advocated by neo-conservative thinkers and their advocates who coalesced around the Bush administration. The geopolitical imagination adopted following 9/11 made the pursuit of more peaceful alternatives less likely, and UK participation in US-led wars more likely. The Church of England provided one of a number of sacred spaces of memory in which the war on terror, as a set of securitized and hyper-militarized discourses and practices in US and UK domestic and foreign policy, was able to materialize.

---

67   G. Ó Tuathail, '"Just Out Looking for a Fight": American Affect and the Invasion of Iraq', *Antipode* 35/5 (2003) 857–801.

68   J. Wallis, *God's Politics: Why the American Right Gets It Wrong and the Left Doesn't Get It* (Oxford: Lion Hudson 2005).

Chapter 5

A 'New Mecca for Terrorism'? Unveiling the 'Second Front' in Southeast Asia

Chih Yuan Woon

Introduction: The Second Afghanistan? The New Mecca of Terrorism?

On 4 May 2005 the Australian commentary program *Dateline* aired a special feature on the Southern Philippines island of Mindanao, espousing on themes at the critical interface of society, politics and counter-terrorism. Amongst the panel of 'experts' interviewed, Acting US Ambassador to the Philippines, Joseph Mussomeli, provided one of the most critical, albeit controversial, overall conspectuses of the situation in Mindanao. Citing classified information pertaining to alliances forged between Philippines Muslim insurgency groups and other extra-local terrorist organizations such as al-Qaeda and Jemaah Islamiyah (JI), the ambassador could only conclude with elegiac assertions:

> Certain portions of Mindanao are so lawless, so porous, the borders, that you run the risk of it becoming like an Afghanistan situation. Mindanao is almost, forgive the poor religious pun, the new Mecca for terrorism.[1]

Edward Said has bequeathed to us the felicitous term 'imaginative geographies', reminding us of the unequal power relations that enable knowledge about other places and peoples to be re-produced.[2] Mussomeli's evocation of Mecca and Afghanistan is perhaps an exemplar of the insidious workings of imaginative geographies. By galvanizing the dominant imaginings of these places with terrorism, they can then be superimposed onto Mindanao so that the contours of violence in the area can be made recognizable and identifiable. However, as Said further cautions, imaginative geographies should not be reduced simply to an internally structured archive of knowledge; rather they are performative insofar that they produce effects. Indeed, deployment of US troops to the Southern Philippines (see Figure 5.1) to combat a local Muslim terrorist group in 2002 is indicative of the performance of imaginative geographies that are captured through Mussomeli's vivid caricature.

---

1   J. Mussomeli, 'Mindanao: Crucible of Terror' (4 May 2005) at <http://news.sbs. com.au/dateline/mindanao__crucible_of_terror_130510>.

2   E. Said, *Orientalism* (London: Routledge & Kegan Paul 1978).

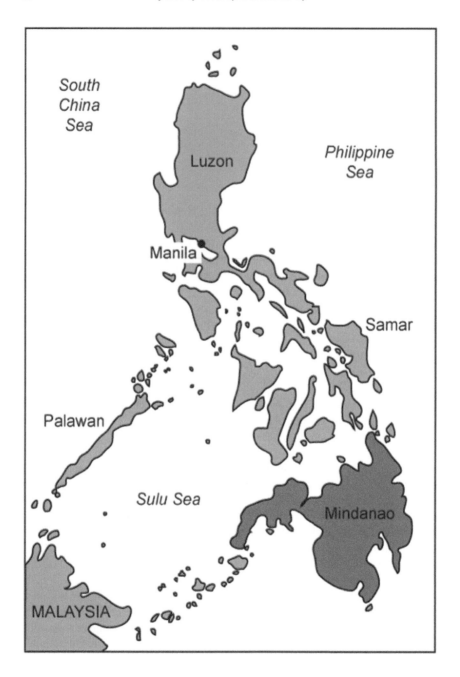

Figure 5.1 Map of the Philippines
Source: Drawn by Jenny Kynaston.

The theme of imaginative geographies and its performative powers are further explored in Gregory's engaging monograph, *The Colonial Present*.[3] Gregory is wary of the plethora of accounts that are quick to use the horrific September 11 attacks on America as a base to launch a dramatic tale of subsequent US military incursions into Iraq and Afghanistan. According to him, relating the story in such a unilinear fashion assumes that there is no relevant prehistory to the occurrences of September 11.[4] Hence rather than seeing these events in isolation, Gregory is attentive to how multiple geographies and histories have condensed and coalesced in ways that have considerably extended the present moment of danger. Premising on the argument that the war on terror draws sustenance from a cultural mobilization, a compelling narrative of three imaginative geographies (comprising of the spatial strategies of locating, opposing and casting out) is thus weaved to draw their intimate connections to the performances of space in Afghanistan, Iraq and Palestine. First, 'locating' mobilizes a technical or techno-cultural register in which opponents are routinely reduced to mute objects in a purely visual field – letter on a map, coordinates on a grid – that produced an abstraction of other people as 'the Other'. In such a formulation, the targets are cities, buildings and infrastructures in Iraq, Palestine and Afghanistan; humans are effaced and somehow invisible from the scene. The second spatial strategy, 'opposing', thrives on a largely cultural register in which antagonism is reduced to a conflict between two binary forces – a unitary and universal Civilization (epitomized by the US) versus multiple, swarming barbarisms that serve as its negation and nemesis (see Chapter 4). The subsequent course of action is clear: all barbarians are to be duly dispatched. Finally, 'casting out' incurs a political-juridical register whereby not only armed opponents (e.g. al-Qaeda terrorists, Taliban troops, Iraqi soldiers) but also civilians and refugees are reduced to the status of *homines sacri*. Their lives did not matter. The sovereign powers of the American, British and Israeli states disavowed or suspended international law so that men, woman and children were made outcasts, placed beyond the protections and affordances of the Modern.

Building on Gregory's provocative insights, this chapter will critically interrogate the geographical extension of the US-initiated war against terrorism to the South East Asian region. Demarcated by the US as the 'second front' in the global war on terror, South East Asia has acquired a notorious reputation for being the 'new theatre' for staging transnational terrorism.[5] The underlying view is that terrorism in this part of the world is intricately linked to the global network of al-Qaeda, which orchestrates however loosely, the training, financing and even operational leadership of local incidents such as the Bali attacks in 2002. Such a conceptualization, I argue, actively employs the spatial strategies of 'locating' and

3 D. Gregory, *The Colonial Present* (Oxford: Blackwell 1994).

4 For similar arguments, see J. Butler, *Precarious Life: The Power of Mourning and Violence* (New York: Verso 2004).

5 R. Gunaratna, *Inside Al-Qaeda: Global Network of Terror* (New York: Berkeley Books 2003).

'opposing', where the threat of an external 'Other' (al-Qaeda) is contextualized within the region. In addition, South East Asian states have aligned themselves (albeit to varying degrees) to America's anti-terrorism agenda in the region where there is a preoccupation with paramilitary interventions and intelligence sharing so as to 'cast out' the so-called 'terrorist' factions. This is hardly surprising given the binarism that Bush proposes in which only two positions are possible – Either you're with us or you're with the terrorists – makes it untenable to carve out a space where one opposes both and queries the terms in which the opposition is framed.[6] This is however not to suggest that the active agencies of these states are circumscribed by the contemporary geopolitical ambitions of the US. As Glassman has noted, the war on terror has also provided an apposite opportunity for the governments in South East Asia to exercise increasing surveillance and control over 'dissident groups' within their national boundaries.[7]

However as this chapter will seek to exemplify, the 'second front' label is counterproductive for 'rooting' out the sources of terror, a utopian end-state envisaged by President Bush. The three spatial strategies are symptomatic responses with scant regard for the underlying *conditions* that harness terrorism. To this end, I echo Gregory in proposing three counter-geographies – contextualizing, connecting and cosmopolitan – so as to tease out the nuances and get to the 'roots' of violence. Revisiting the case of the Philippines, I seek to *contextualize* conflicts and 'terrorism' within the country in order to go beyond any simplistic suggestions that Philippines is merely a duplicate of Afghanistan, a safe haven for Islamic terrorism. In particular I seek to draw the *connections* to the multiple histories and geographies that mingle in myriad ways to contribute to the spatial manifestations of contemporary violence in the Philippines. This will allow for a critical rethinking of the fundamental agencies that are complicit in creating the conditions for violence to breed. The conclusion will reflect on the potentialities of the notion of cosmopolitanism as an exhortatory ideal to intervene for a more humane and peaceful world.

The 'Second Front': Politics, Negotiations and Responses

If there were any reductionist thoughts of the September 11 attacks as signals for the prophetic coming of the 'clash of civilizations'[8] that involved only two coherent entities (the 'West' and the 'Muslim World'), they were quickly complicated by the discovery of the terrorist plot aimed at Singapore in December 2001. The initial relief that the Singapore authorities had managed to thwart a series of malicious plots targeted at local infrastructures (for example water plants, train stations)

---

6   G. Bush, 'Address to a Joint Session of Congress and the American People' (20 September 2001) at <http://www.whitehouse.gov/news/releases/2001/09/20010920-8.html>.

7   J. Glassman, 'The "War on Terrorism" comes to Southeast Asia', *Journal of Contemporary Asia* 35/1 (2005) 3–28.

8   S. Huntington, 'The Clash of Civilizations? The Next Pattern of Conflict', *Foreign Affairs* 72/3 (1993) 22–50.

and US military installations within the country was soon overwhelmed by the macabre revelations of the intra-regional dimension of this challenge. In a white paper published by the Singapore Ministry of Home Affairs, it was reported that 35 people were arrested as perpetrators of the planned attacks and that most of them have allegiance to the radical Islamic group JI, known to be active in South East Asia.[9] The paper points out that these plots are linked to the wider objective of JI to create a pan-South East Asian Islamic state comprising the Muslim-majority areas of southern Philippines, Indonesia, Malaysia, Singapore and southern Thailand. Laying out 'evidence' collated from different sources which include a pan-regional blueprint and private letters/emails, elaborate efforts were spent on enumerating the connections of JI with al-Qaeda in particular and how JI has benefited from trans-regional training and support (see Figure 5.2). The conclusion is thus clear: al-Qaeda and its network of adherents represent an even larger threat to South East Asian security by transcending local or national grievances and faultlines.

The findings of the white paper confirm the suspicions of many that South East Asia is indeed a thriving terrorism hotbed, with its deserved status as the 'second front' in the global war on terror.[10] Since the September 11 attacks on America, transnational terrorism has often been touted as the worst nightmare in the security imaginings of the region's governments. There is widespread belief that with its defeat in Afghanistan, al-Qaeda elements have shifted their attention to South East Asia. More importantly, there are simultaneous migratory flows of South East Asians training in Afghanistan who could respond to the al-Qaeda leadership's periodic call for terrorist strikes, especially against entertainment spots frequented by Western tourists (for example Bali and Phuket). Hence, the underlying view is that South East Asia functions as a potent regional tentacle for al-Qaeda to extend its influence and reach due to a plethora of factors: multiethnic societies; weak and corrupted regimes with tenuous control over peripheral areas; ongoing separatist insurgencies that lend themselves to exploitation by foreign elements; and newly created 'democratic' space in some of its larger polities such as Indonesia and the Philippines which have found it difficult to mobilize public support for security regulations to ensure preventive suppression of terrorist factions.[11]

---

9   Singapore Ministry of Home Affairs, *White Paper: The Jemaah Islamiyah Arrests and the Threat of Terrorism* (January 2003) at <http://www.mha.gov.sg/publication_details. aspx?pageid=35&cid=354>.

10   J. Gershman, 'Is Southeast Asia the Second Front?', *Foreign Affairs* 81/4 (2002) 60–74; A. Tan, 'Southeast Asia as the "Second Front" in the War Against Terrorism: Evaluating Threats and Responses', *Terrorism and Political Violence* 15/2 (2003) 112–38; A. Rabasa, 'Southeast Asia: The Second Front?', *Asia Program Special Report* (2003) at <http://wilsoncenter.org/topics/pubs/Asia%20Report%20112.pdf>; D. Wright-Neville, 'Prospects Dim: Counter-terrorism Cooperation in Southeast Asia', *Asia Program Special Report* (2003) at <http://wilsoncenter.org/topics/pubs/Asia%20Report%20112.pdf>.

11   A. Acharya, 'Deadly Discourse: Reflections on Terrorism and Security in an Age of Fear', in M. Vicziany (ed.), *Controlling Arms and Terror in the Asia Pacific* (Cheltenham, UK: Edward Elgar 2007) pp. 83–96.

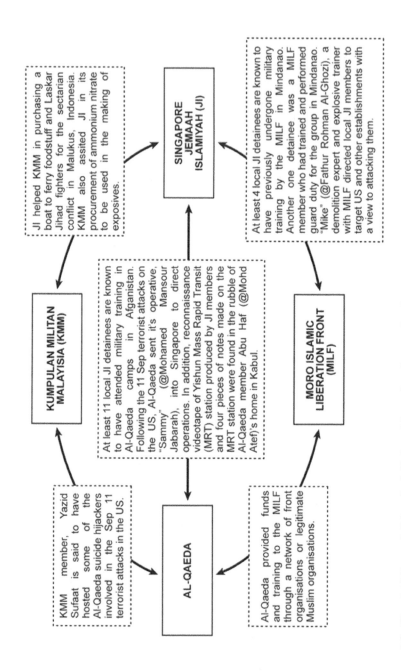

Figure 5.2 Chart of Jemaah Islamiyah links with regional organizations
Source: Adapted from Singapore Ministry of Home Affairs (2003, p. 9), by Jenny Kynaston.

With the rhetoric of the 'second front', US foreign policy has garnered forceful initiatives for a revived engagement with South East Asia. As Glassman has pointed out, the post September 11 events have allowed the US to regain the position of hegemony in the region, a privilege relinquished since the end of the Vietnam War and the loss of rights to the Clark Air Base and Subic Bay Naval Base in the Philippines in the early 1990s. Associating transnational terrorism with South East Asia via the label of the 'second front' indeed provides legitimating reasons for US ventures into the region.[12] After all, such endeavours are justified in the name of self-defence (pre-empt further terrorist incursions on America) and for the greater good (rooting out terrorism for the betterment of the global community). However, while such justifications for gaining access to the region may appear sound from the US perspective, it does not necessarily translate into South-east Asian states aligning themselves to such views, thereby lending their support and cooperation to America. Thus there is a need to analyse the discursive frame of the 'second front' in tandem with another widely circulated rhetoric – President Bush's assertion that 'Either you are with us or you are with the terrorists.'[13]

Butler has argued that this bifurcation 'returns us to an anachronistic division between 'East' and 'West' and which in its sloshy metonymy, returns us to the invidious distinction between civilization (our own) and barbarism (now coded as Islam itself)'.[14] Such ideological bloodlines signify that to oppose America's global war on terror is tantamount to condoning barbaric acts intrinsically related to Islamic terrorism. The possibility of a third space, which allows the nourishment of resistances to both America's war initiatives and terrorism, is thus foreclosed. Given such circumstances, countries in the region have been quick to express sympathies for and condemn the terrorist attacks on America. However this unison has not led to a homogeneous interpretation of the extent of danger posed by terrorism to national security and regional order. It must be noted that the *discursive* responses to the idea of the 'second front' are mediated through the domestic politics of the various Southeast Asian states.[15] Yet interestingly, such varied conceptualizations of terrorism in the region have not produced distinctively heterogeneous *policies* to deal with the issue. Indeed, as the ensuing discussion will reveal, the three spatial strategies of locating, opposing and casting out are central responses of the Southeast Asian states. There is an intense desire to locate terror, represent it as the threatening 'Other' so that measures can be undertaken for its eradication. The overwhelming obsession with policing and paramilitary measures, I suggest, has downplayed the complex permutations of political conditions and socio-cultural dynamics that harness terrorism.

---

12   Glassman (note 8).

13   Bush (note 7).

14   Butler (note 5).

15   D. Capie, 'Between a Hegemon and a Hard Place: The War on Terror and Southeast Asian–US Relations', *The Pacific Review* 17/2 (2004) 223–48; K. Nathan, 'South-East Asian Responses to Arms and Terror', in M. Vicziany (note 13) 159–79.

For instance, Singapore maintains an uncompromising stance on the general issue of terrorism and its non-secular (Islamic) character. As the Singapore High Commissioner to Australia, Mr Eddie Teo puts it, 'It may not be politically correct to focus on the relationship between Islam and terrorism. However ... what [JI members] were ... taught was that to be a good genuine Muslim, you would have to have the West, bring down secular pro-Western governments in the region and pave the way for an Islamic regional government.'[16] The government of Singapore has made it clear that since the country is heavily dependent on foreign trade and tourism, the impact of a terrorist attack could be catastrophic for the economy. Furthermore, terrorism also has the potential to tear apart Singapore's multiracial and multi-religious social fabric. As a result, Singapore has openly allied itself with the West particularly the United States, not only on business and commercial ventures but also on global security and strategic issues. On the other hand, Indonesia and Malaysia are less inclined in acknowledging terrorism as an existential threat (as Singapore does) and they are wary about drawing the connections between terrorism and Islam. Such attitudes have to be understood in light of the large Muslim population in these two countries. There is an ambivalent, if not sometimes hostile perception of the policies of the US, which is seen to be anti-Islam and oppressive of Muslims worldwide. Furthermore, there is the view that declaring a war on terrorism will not yield positive outcomes since the roots of the 'problem' are given scant attention. As such, Indonesia and Malaysia have been highly reserved in linking political and religious organizations operating within their territory to the global al-Qaeda network. Despite their discordance with the US on the conceptualization of terrorism, it is interesting to note that Malaysia and Indonesia have not repudiated defence or security cooperation with Washington. While they are reticent about allowing American direct presence in the region for anti-terrorism purposes, they are however keen on intelligence exchanges and receiving monetary and technical aid from US in order to enhance their terrorism combatant abilities.[17] There are no signs of a commitment to get beyond the superficial manifestations of terrorism. A critical disjuncture thus arises at the discursive and operational level for Malaysia and Indonesia – acknowledging the need to address the roots of terrorism yet following US in its symptomatic counter-terrorism strategies at the ground level.

The decision of South East Asian states to engage bilateral relations with the US on terrorism issues (instead of a collective ASEAN-US interaction) is

---

16   W. Teo, 'The Emergence of Jemaah Islamiyah in Singapore', Speech at Brookings Institution (25 November 2002).

17   In August 2002, the US State Department announced plans to give Indonesia US$60million worth of counter-terrorism assistance. This include US$47million to strengthen the capacity of the police and US$16million for technical assistance. On the other hand, a US–Malaysia Anti-terrorism Pact was signed in Washington in May 2002 which outlined plans for cooperation in the areas of defence, banking, intelligence sharing, border control and law enforcement.

indicative of the uneasy tensions that underpin regional anti-terrorism efforts. The lack of a unifying legal-politico perspective on terrorism in the region meant that negotiations and compromises have to be undertaken. Indeed at the discursive level, ASEAN proclaims to adopt a comprehensive twin approach of using military means to counter the threats of terrorism while simultaneously addressing the often deep-seated origins of these violent endeavours. This approach satisfies and attends to the divergent views of terrorism in South East Asia. However, as with the case of Indonesia and Malaysia, the official rhetoric does not coincide with the enacted policies. For example, in 2007, ASEAN members signed the ASEAN Convention on Counter-terrorism to provide a framework for deeper regional cooperation to counter, prevent and suppress terrorism.[18] Provisions for rapid information sharing, and the establishment of a common database as well as standardized procedures for prosecution and extradition make the convention significantly different from some of the earlier regional initiatives. The legally binding nature of the convention is surprisingly ambitious, given ASEAN's usual practice of embarking on non-legalistic forms of cooperation. ASEAN's initiatives to deal with terrorism, however, have included little in terms of concrete counter-terrorism mechanisms. As Acharya and Acharya have argued, the long-simmering tensions between various parties (e.g. Singapore with Malaysia, Indonesia and Philippines) have resulted in cracks and fissures in this security coalition, hampering the effective exchange of sensitive counter-terrorism information.[19] Once again, there are no signs in such regional treaties of the desire to tackle the causes of terrorism. There is only a singular pathway that reactions to terrorism can tread on and advance. This I suggest is a result of the principle of non-interference that ASEAN member states adhere to.[20] Since conditions fuelling terrorism differ from country to country, they are detached from regional responsibilities, falling under the domain of domestic politics. Framed in such a manner, the genesis and nuances of the phenomena of terrorism needs to be recovered.

(Dis)Illusions Associated with the 'Second Front'

As briefly alluded to in the preceding discussion, skepticism and resistances have been soaring high in some ASEAN countries about the latest US interest in the region. Such sentiments offer a glimpse into the problematic nature of how the US imagines South East Asia. By engaging in a deconstructive reading of the 'second front' label that is being imposed on South East Asia, I seek to illuminate

---

18   For more information about the ASEAN Convention on Counter-terrorism, refer to the ASEAN website at <http://www.aseansec.org/19250.htm>.

19   A. Acharya and A. Acharya, 'The Myth of the Second Front: Localizing the War on Terror in Southeast Asia, *The Washington Quarterly* 30/4 (2007) 75–90.

20   ASEAN has implemented the non-interference policy in member states' internal affairs since its founding in 1967.

how such branding mystifies the whole idea of terrorism in the region, making it problematic and counterproductive for US's penultimate aim of rooting out the sources of terror.

First, the use of the term 'second front' implies a conceptual and operational link between terrorist groups in South East Asia and the Middle East. As outlined earlier, the international dimensions of terrorist activity in South East Asia particularly al-Qaeda's direct or indirect presence have been (re)emphasized. Indeed many terrorism 'experts' with advisory roles to the policy-makers in America are playing an integral role in informing and reinforcing these connections. They, more often than not, focus on the who and what of terrorist activity in South East Asia, having at their core highly descriptive compilations of dates and meetings involving terrorist or terrorist related individuals or groups. Despite an overwhelming emphasis on details, their analyses suffer from vagueness in defining terrorism and largely fail to elucidate the significances of the event and linkages they describe. Zachary Abuza, for example, not only draws the nodes linking JI and al-Qaeda but also lists the meetings and personal connections that bring two Philippine groups, Abu Sayaf Group (ASG) and the Moro Islamic Liberation Front (MILF) into contact with al-Qaeda.[21] The underlying impetus for tackling this task is however unclear. On one hand he describes MILF's condemnation of terrorism and active involvement in political negotiations. Moreover he judges the assertion that the ASG is linked to al-Qaeda to be 'tenuous at best'.[22] On the other hand, he sends out contradictory signals – MILF and ASG are discussed under the heading of 'From Parochial Jihadis to International Terrorists' and the book frequently stresses the 'deep links' between al-Qaeda and 'home grown Muslim insurgencies such as MILF and ASG'.[23]

Another influential study tries to unpack the notion of 'terrorist-linked', referring it to a group's aspiration to go beyond its political and/or military struggle for separatism, banditry or some other cause.[24] The basic argument is that the way in which this threat is escalated seems to be that terrorist links – personal, material or ideological – serve as reservoir of support for the terrorist activities of groups such as al-Qaeda. Gunaratna includes as part of al-Qaeda's 'network' a group referred to as Kumpulan Mujahidin Malaysia (KMM), along with JI, ASG and MILF. The most explicit that he ever gets to defining what being part of the networks connotes is that while KMM is an independent group, 'Al-Qaeda works via KMM'.[25] Gunaratna also warns of the serious threats that radical groups pose

---

21   Z. Abuza, *Militant Islam in Southeast Asia: Crucibles of Terror* (Boulder, CO: Lynne Reinner 2003). Abuza is an acclaimed terrorism expert and a Senior Fellow at the United States Institute of Peace (USIP). USIP is funded by the US Congress to give policy advice to resolve and prevent violent international conflicts.

22   Ibid., p. 99 and p. 112.

23   Ibid., p. 108.

24   Gunaratna (note 6).

25   Ibid., p. 261.

to countries like Indonesia, with the support function clearly underlined.[26] In some cases, the linked groups in his narrative are lawful organizations which simply have some kind of Islamic identity. He makes no attempt to either elaborate on what these links consist of or assess their meanings. Hence, due to the inherent weaknesses in these analyses, many Southeast Asians view such conceptions as unduly alarmist and misrepresentative of terrorism in the region as having global origins.[27] As such, a blindspot obscures local agendas, historic grievances and experiences of marginalization at the national or subnational scale.

While the idea of the 'second front' does implicate external factors in accounting for terrorism in South East Asia, it also proffers the crucial question of why such alliances between Al-Qaeda and regional terrorist groups are forged in the first place. After all, meetings and linkages between people/groups do not automatically translate into cooperative actions. This is where another clause inherent in the 'second front' designation comes into play: the transfer of *ideology* from al-Qaeda to other groups. There is almost an espousing of a trickle-down effect where al-Qaeda's radical version of Islam is transposed to the various groups/individuals in South East Asia. The path leading to religious fundamentalism is often discussed but such trajectories tend to be obscurely sketched. There is however invariable belief that *deviancy* is the underlying mechanism that is at work in inducing radicalization. For instance, a report prepared for US intelligence by the Federal Research Division gives a thorough dissection of the psychological profiles of terrorists.[28] It argues that depending on the type, aims and even gender/nationality of the terrorist, varying traits/conditions such as mental illness, social alienation and fanaticism can be found on her/him. Similarly, popular/academic works on terrorism in South East Asia help to perpetuate such a stance. Millard's recent book is a good representation of this sort of research.[29] According to him, Wahabi thinking is proliferated via Saudi-funded missionizing, which had few problems in finding a receptive audience in South East Asia due to the envy of the West and anger at political repression. Terrorists who are produced in this way are afflicted by 'madness' which makes them 'cross the line that divides sanity from pathology'.[30] Their critical thinking and judgmental abilities are clouded, resulting

---

26   Ibid., pp. 286–70.

27   F. Noor, 'Globalization, Resistance and the Discursive Politics of Terror Post-September 11', in A. Tan and K. Ramakrishna (eds), *The New Terrorism: Anatomy, Trends and Counter-Strategies* (Singapore: Eastern Universities Press 2002) 157–77; J. Putzel, 'Political Islam in Southeast Asia and the US–Philippine Alliance', in M. Buckle and R. Fawn (eds), *Global Responses to Terrorism: 9/11, Afghanistan and Beyond* (London: Routledge 2003) 176–87.

28   Federal Research Division, *The Sociology and Psychology of Terrorism: Who Becomes a Terrorist and Why?* (1999) at <http://www.loc.gov/rr/frd/pdf-files/Soc_Psych_of_Terrorism.pdf>.

29   M. Millard, *Jihad in Paradise: Islam and Politics in Southeast Asia* (Armonk, NY: ME Sharpe 2004)

30   Ibid., p. 31 and p. 51.

in an acceptance of the 'convolutions of reason' of clerics like Basyir.[31] Ultimately terrorism is seen as a mental disease inflicted on them by demonic preachers: at the centre of each web of terrorists is 'a militant cleric bringing believers under his spell, convincing them to hate outsiders'. Such Islamists are 'at the edge of psychosis'.[32] This interpretation is colourful but otherwise not untypical of religion-driven accounts. As Ramakrishna has espoused, an Islamic terrorist is the product of indoctrination, individual pathology and envy of the West leading to a rejection of modernity itself.[33]

Given such a formulation, the deployment of Islam as an emotive force to mobilize violence is a distortion of orthodox religious teachings. It comes as little surprise then that there is strong impetus in the region to construct 'Southeast Asian Islam' as being 'moderate'. Religious leaders and Islamic scholars of Indonesia's largest Muslim organization, Nahdlatul Ulama (NU), for example, has affirmed to then US Secretary of State Colin Powell shortly after the September 11 attacks that 'Islamic fanaticism, extremism or fundamentalism is not a popular stance among NU members'. Their penultimate goal is to 'sustain and promote moderate thinking and a pluralist and tolerant attitude among the younger generation' so that Indonesia can embark on a 'moderate path in developing Islamist thought'.[34] As can be witnessed from such exhortations, attention to the propagation to radical Islam is frequently matched with recommendations to adopt an 'ideological' response – to combat the spread of terrorist behaviour or militant sympathies by disseminating 'good' Islamic teachings. Singapore is exemplary of this approach, given its emphasis on 'orthodox' Islamic education for Muslims and enlisting the support of 'moderate' religious leaders.[35]

With the somewhat excessive bombardment of ideas related to radical Islam and terrorism, certain dominant albeit problematic imaginings are being transfused. While most are careful to point out that South East Asian Islam is generally peaceful and moderate, they however paradoxically contribute to the idea that Islam is potentially threatening by analysing militancy and terrorism through the religious lens. Categories such as moderate, fundamentalist, militant and terrorist are sometimes presented as possible progressive stages through which individuals may move.[36] A second problem with the categories of 'moderate' and

31 Ibid., pp. 52–3.
32 Ibid., p. 68 and p. 99.
33 K. Ramakrishna, *'Constructing' the Jemaah Islamiyah Terrorist: A Preliminary Enquiry*, Working Paper No. 71 (Singapore: Institute of Defence and Strategic Studies 2004).
34 *Asia Times Online*, 'Southeast Asia Still Islam's Moderate Face' (31 July 2002) at <http://www.atimes.com/atimes/Southeast_Asia/DG31Ae02.html>.
35 S. Kadir, 'Mapping Muslim Politics in Southeast Asia After September 11', *The Pacific Review* 17/2 (2004) 199–222; E. Tan, 'Reaching Out to the Moderate Muslims in Singapore: The Role of Soft Law, Norms, Education and Dialogue', Paper Presented at the Annual Meeting of the Law and Society Association (July 2006).
36 Ramakrishna (note 36); D. Wright-Neville, 'Dangerous Dynamics: Activists, Militants and Terrorists in Southeast Asia', *The Pacific Review* 17/1 (2004) 27–46.

'radical' is that they reproduce a dichotomy that echoes the 'with us' or 'against us' rhetoric so prevalent in the war on terror. An inter-related chain of binaries is set up between 'moderate' Islam which is benign, peaceful and not antagonistic and 'extreme' or 'fundamentalist' Islam which is hostile to the US and prone to violence. The litmus test for 'moderate' Muslims thus becomes attitude to the US. Using the categories of moderate or radical Islam also underplays grievances in a wider geopolitical context such as US attacks on Iraq and Afghanistan. Because such attitudes are often referred to as being an adjunct of a religious disposition, their political content (and their potentially universal rationality) is denied. Noor has noted that Osama Bin Laden and the Taliban have come to mean anything and everything anti-American and for this reason, 'wearing Osama or Taliban T-shirts does not make one a *jihad* terrorist any more than wearing a pair of Nike shoes makes one a committed American capitalist'.[37] But the bewildering thing is that when it comes to Muslims expressing anti-American views, there is the active employment of the categories of moderate versus radical Islam, thus portraying an underlying religious motivation at work.

Third, labeling South East Asia as a second front projects the US led war on terror as a pretext for Washington to entrench its military supremacy globally and regionally while asserting its narrow economic and strategic interests at the expense of regional security and welfare. In South East Asia, both mainstream and secular nationalist groups view the US role with serious misgivings. This arguably has to be attributed to the long history of US interventions in the Middle East, which is widely regarded by South East Asians as pro-Israel and simultaneously marginalizing of Palestinian Muslims' interests. As Surin Pitsuwan, former Foreign Minister of Thailand, explains, a strong sense of 'primordial' resentment exists among all 'Muslims around the world, particularly here in Southeast Asia' rooted in the belief that their sentiments around Jerusalem, which after Mecca and Medina is the third-holiest site in Islam, 'have never been seriously accommodated' by Washington.[38] Furthermore, the unilateral US decision to invade Iraq has fuelled this resentment. Malaysia's former Prime Minister, Mahathir Mohammad goes so far as to claim that Iraq's occupation by American forces signifies 'an increasingly violent culture' which may well be a 'prelude to the Fourth World War'.[39] The complex geopolitical conflicts in Iraq have demonstrably dampened popular support for the war on terror and inflamed Muslim anger in South East Asia. In Indonesia alone, for example, the favourability rating for the US fell from 61 per cent to 15 per cent, according to a survey in June 2003.[40] President Bush did not

---

37 Noor (note 30).

38 S. Pitsuwan, 'Strategic Challenges Facing Islam in Southeast Asia', Lecture in Singapore (5 November 2001).

39 M. Mohammad, 'US Policy in Iraq, Iran Prelude to 4th World War' (8 June 2006) at <http://english.peopledaily.com.cn/200606/08/eng20060608_272051.html>.

40 *The Straits Times*, 'Bush's Asia Trip: More Than Just Trade, Terror and Thank Yous' (18 October 2003).

receive a warm reception from the Muslim clerics when he visited Indonesia in 2003. He was told blatantly that US foreign policy's insinuations of Muslims as terrorists have been harmful to the global Muslim community and was one of the root causes of terrorist attacks targeting US interests.[41] As this simple illustration exemplifies, by pursuing a wayward military solution in Iraq and elsewhere, the US perpetrates and displays its own violence, offering a breeding ground for new waves of young Muslims to join terrorist organizations: 'Ignoring its images as the hated enemy for many in the region, the US has effectively responded to the violence done against it by consolidating its reputation as a militaristic power with no respect for lives outside the First World.'[42] Responding with more violence only presents 'further proof' that the US has violent and anti-sovereign designs on the region. Consequently, governments in the region found it difficult to engage in an open security partnership with the US out of fear of domestic criticisms.

From the preceding critiques of the 'second front' rhetoric, it can be seen that the label assumes that an individual subject (al-Qaeda, Osama Bin Laden, the pathological South East Asian terrorist etc.) is the first link in a causal chain that forms the meaning of accountability. To take self-generated acts of the individual as the point of departure becomes the occasion for casual reduction, foreclosing the possibility of engaging in other critical questionings: What kind of environments shape the rise of such individuals? What kind of social conditions help to form the ways that choice and deliberation proceed? Against what kinds of conditions of violation do they respond? Situating individual responsibility in light of its collective conditions undoubtedly opens up a whole series of epistemological insights and provides a grid of intelligibility that current meta-narratives on terrorism shut down. In the absence of these other storylines, South East Asian states (subtle or overt) support for US ambitions in this region will continue to manifest themselves as a flawed singular approach, focusing on the spatial strategies of locating, opposing and casting out. By identifying and targeting individual 'terrorists' and their links, these strategies do nothing to ameliorate the social conditions that enable the continuous breeding of 'new terrorists'. Hence, dissenting from the 'second front' in the global war on terror is not to dismiss terrorism as a highly dangerous threat to security and well-being; it is to reject a tendency to conflate all sorts of challenges under an overarching framework and to encourage a search for alternative understandings of violence and responses to it. Indeed, the Philippines case provides a good example to interrogate the aforementioned issues, demonstrating the complexities of 'terrorism'. As will be discussed in the following section, contextual geographies, with attention attuned to the critical histories and geographies that harness terrorism, have immense value in allowing other storylines related to the war on terror to be heard (and seen).

---

41   A. Acharya, 'APEC Summit: Regionalizing the War on Terror' (2003) at <http://www.pvtr.org/pdf/RegionalAnalysis/AsiaPacific/War%20on%20Terror.pdf>.

42   Butler (note 5).

Geo-historical Inquiries into Muslims' Struggles for Self Determination in Southern Philippines

The case of the Philippines deserves closer scrutiny not only because of the country's strong endorsement of the 'second front' in the war on terror but also due to the government's unprecedented move of including American troops in its domestic operation against local 'terrorist groups'. Manila was quick to latch on to the anti-terror bandwagon in September 2001 and this pledge of allegiance to the US cause has resulted in the country being the largest beneficiary of American largesse in South East Asia. President Gloria Arroyo visited Washington shortly after the September 11 attacks, declaring unrelenting support for US actions and that 'the American and Filipino people stand together in the global campaign against terrorism'.[43] This strong public exhortation reaped extensive benefits for the Philippines military which obtained more than US$100 million in financing and equipment. In addition, the Philippines economy received a huge boost – a promised package including trade credits, tariff reductions and debt write-offs, potentially worth more than US$1 billion from the US.[44] Not denying the significance of these monetary rewards, the event that perhaps directed kaleidoscopic interest in the Philippines (and arguably South East Asia) is the direct involvement of US forces in eradicating terrorist factions in the Philippines. In January 2002, the Bush Administration made a major decision to send some 1200 US troops to the Southern Philippines to fight the Abu Sayaf Group (ASG) operating around the island of Basilan.[45] As Gershman purports, the 'second front' in the war on terror has been transformed, at this point, from a discursive terrain to a set of material practices with alarming consequences.[46] American and Filipino forces went to great lengths to emphasize that the troop deployment and operations are merely 'training exercises', but the integral part played by US Special Forces in a number of anti-terrorism operations in June 2002 suggests otherwise. This joint effort, publicly known as *Balikatan* (shoulder-to-shoulder), hints at the close security relationship between the US and the Philippines and it did not surprise many when the latter was made a non-NATO ally of the US soon after in 2003.

The seemingly coherent US–Philippines security partnership can be disrupted if we look beyond a(n) (inter)state level of analysis. A multi-scalar approach will enable the revelation of the nuances and limitations of such an alliance when evaluated alongside regional and local politics. This helps to debunk the

---

43   The White House, 'Joint Statement between the United States of America and the Republic of the Philippines' (20 November 2001) at <http://www.whitehouse.gov/news/releases/2001/11/20011120–13.html>.

44   Ibid.

45   J. Perlez, 'Philippine Troops Eagerly Await US Help and Arms', *The New York Times* (12 February 2002).

46   J. Gershman, 'Is Southeast Asia the Second Front?', *Foreign Affairs* 81/4 (2002) 60–74.

common myth that the state has absolute power in formulating 'counter-terrorism' policies without considering the performative abilities of other actors at different geographical scales in this entire process. Within the Philippines, there have been extensive criticisms of the 'patron' role of the US by some Filipino nationalist politicians, non-governmental organizations and the media who see the dangers of growing dependence on the Americans in dealing with an essentially home-grown terrorist threat. In particular, *Balikatan* has been criticized as unconstitutional and undermining the sovereignty of the Philippines. Opposition leaders have also questioned the legality of US troops deployment in Mindanao and the motive for holding joint 'military exercises' in that part of the country, as they worry about the Philippines becoming a second Vietnam, especially if US forces are targeted.[47] On the other hand, at the regional level, the support of the Philippines for the US campaign has to be weighed against the collective ASEAN position on this issue. As mentioned earlier, the ASEAN exhortatory stance diverges from the highly military-oriented approach of the US, advocating a more holistic treatment of terrorism in the region. The Philippines has somewhat gone along with this collective position of ASEAN – while recognizing the need for US military assistance in containing the threat posed by radical groups predominantly in the Southern Philippines, the Arroyo Administration has repeatedly underscored the socioeconomic roots of terrorism in the country. To some extent, the need to anchor the Philippines' policy with ASEAN's position also serves a strategic purpose. Given the uneasy domestic tensions arising from (direct) US interventions, the Philippine government is wary of being judged as simply a mere shadow of US unilateralist initiatives in the region. Hence its membership of ASEAN offers it some respite, serving as a 'balancer' in this regard.

The complex negotiations on the ground underscore the continuing hostility to any US military presence on the part of many Filipinos.[48] Despite these challenges, Glassman contends that the ability of the US to foster a working relationship with the Philippines is a consequence of the dynamic interaction between the geopolitical and economic ambitions of the elites in these two countries.[49] He argues that US neoconservatives had already targeted South East Asia as a key site for the build-up of military forces even before the September 11 attacks on America. This emphasis on South East Asia has nothing to do with the presumption of a threat to US security from Islamic terrorist organizations; rather it had everything to do with the goal of attempting to re-assert US hegemony in a part of the world conceived as home to China – the next 'great-power' threat to US hegemony and one of the world's most dynamic and promising regional economies. Hence the 'second front' discourse provides a convenient pretext for the expansion of operations already

47 N. Morada, 'Progress and Setbacks in Philippine–US Security Relations', in Vicziany (note 13) 181–93.

48 D. Pastrana, 'Setback for US Plans to Send Combat Troops to the Philippines' (14 March 2003) at <http://www.wsws.org/articles/2003/mar2003/phil-m14.shtml>.

49 Glassman (note 8).

stated as policy objectives across the spectrum of Washington elites. However the fact that the US military has been able to state forthrightly its desire for renewed access to the Philippines cannot be divorced from US government's continued influence in Manila due to (neo)colonial legacies (the Philippines being a former colony of the US following the 1898 Spanish-American Wars), as well as ongoing dependence of the Philippine elite on foreign direct investment from the West and access to US markets. The scene is set for the Philippine elites who have 'battled against widespread nationalist sentiment with the opportunity to sell their own collaboration with the US government as a project in the national interest'.[50]

As a consequence, Glassman contends that there is a *reversal* of processes of democratization and de-militarization in the Philippines. Indeed, the logic underpinning counter-terrorism policies within the Philippines attests to this incisive observation by Glassman. Shortly after the ascription of the 'second front' label to South East Asia, Philippines adopted a 14-point program that focuses on law enforcement on terrorism. The initiatives include establishing a Cabinet Oversight Committee to set strategy, designating the national security adviser to coordinate intelligence exchanges with foreign counterparts, tasking the Securities and Exchange Commission to conduct an inventory of organizations that might be involved in terrorist financing and synchronizing domestic with global counter-terrorism efforts.[51] It is not difficult to decipher that the attention is once again on the tri-spatial strategies of locating, opposing and casting out. In addition, there are ongoing debates about whether to pass the newly devised Anti-Terrorism Bill. In essence, this bill does not deviate much from the fundamental aims of the 14-point program, delineating what falls under terrorism, putting institutional mechanisms in place to prevent and suppress it and providing penalties for terrorism and related purposes. However what is innovative (and controversial) about this Bill is the presence of systemic apparatuses to avoid guarantees associated with the criminal process. In other words, the Anti-Terrorism Council is afforded unbridled discretion in defining and prosecuting acts of terrorism. By invoking this administrative process, the Philippine government is avoiding the normal guarantees associated with criminal process in much the same way as the military tribunals of detainees in Guantánamo Bay have attempted to insulate the executive of the US government and their national security cases from scrutiny by the courts.[52]

As with Malaysia and Indonesia, the Filipino case presents a situation of mismatch between rhetoric and policy interventions. While the Arroyo administration exhorts for a more comprehensive approach to terrorism, the legislations reflect parochial interests, concurring with US-inspired conceptions

---

50   Ibid.

51   C. Hernandez, 'Fighting Terrorism in Southeast Asia: A View from the Philippines', Asia Program Special Report (2003) at <http://www.wilsoncenter.org/topics/pubs/Asia%20Report%20112.pdf>.

52   C. Donnelly, 'Counter-Terrorism Legislation in the Philippines', in Vicziany (note 13) 194–216.

of terrorism. The overwhelming concern is thus not in addressing the socio-economic roots of terrorism; rather debates centre on which groups have links to foreign terrorist organizations and what should be done to disable their functional capabilities. For instance, post-September 11 claims emanating from the US and Philippine government point to the ASG as a Philippine extension of al-Qaeda. ASG's heightened prominence on the terrorism radar is perplexing, given that President Arroyo has characterized ASG as no more than 'a money crazed gang of criminals' just prior to the September 11 events.[53] Indeed, on the Philippine Left, the ASG is widely suspected to have been a creation of the CIA and/or Philippine intelligence, formed to splinter and undermine the Muslim insurgency in the Southern part of the country. There have been accusations that rather than receiving support from al-Qaeda, ASG leaders are collaborating with the Philippine military which is willing to condone the group's operations in order to receive more military assistance in the name of engaging with the war on terror.[54] Similarly, relentless efforts are undertaken in trying to determine whether MILF are connected to terrorist activities under the directives/supervision of al-Qaeda and JI. The Philippines National Police and the country's military intelligence are adamant that MILF supported JI in operations such as the bombings of Davao International Airport in 2003. However, commissions that have been established to investigate such events exonerated MILF from such a role. MILF has repeatedly renounced terrorism as a means to attaining political ends and President Arroyo is unwilling to include MILF in the list of terrorist organizations for fear of jeopardizing ongoing peace talks with the group.[55] My point in highlighting these messy politics of terrorism is that tremendous efforts are exhausted in debates that may well yield no substantive conclusions, given that it is highly difficult to prove or disprove the claims by various parties. Hence the opportunity is missed for harnessing such efforts into the achievable target of getting to the roots of terrorism, which would have lasting tangible benefits for alleviating violence in the country.

When President Arroyo remarks that 'the best breeding ground [for terrorism] is poverty', she is in fact engaging in an alternative narrative, highlighting the *conditions* not causes. To quote Butler at length:

---

53 International Peace Mission, *Basilan: The Next Afghanistan?* Report of the International Peace Mission to Basilan, Philippines 23–27 March 2002 (2002) at <www.tni.org/reports/asia/basilan.pdf>.

54 N. Klein, 'Stark Message of the Mutiny: Is the Philippine Government Bombing its Own People for Dollars?', *The Guardian* (16 August 2003); J. Roberts, 'Philippine Preseident Renews her Pledge of Loyalty in Washington' (28 May 2003) at <http://www.wsws.org/articles/2003/may2003/phil-m28.shtml>.

55 International Crisis Group (ICG), *Southern Philippines Backgrounder: Terrorism and the Peace Process* (2004) at <http://www.crisisgroup.org/home/index.cfm?id=2863&l=1>.

A condition of terrorism can be necessary or sufficient. If it is necessary, it is a state of affairs without which terrorism cannot take hold, one that terrorism requires. If it is sufficient, its presence is enough for terrorism to take place. Conditions do not "act" in the way that individual agents do, but no agent acts without them. They are presupposed in what we do, but it would be a mistake to personify them as if they acted in place of us.[56]

This short excerpt accentuates the importance of conditions in facilitating and giving rise to terrorist endeavours. It forms the basis of the searching question of *how* terrorism came about, and not *who* are the perpetrators. Hence rather than simply *locating* the 'terrorist' threats, there is a need for the development of contextual geographies that affirm the materiality and corporeality of places and attend to the voices (and the silences) of those who inhabit them. This forces us to duly interrogate what Said and Gregory term contrapuntal geographies, the complex webs of (historical and geographical) connections that contribute to the current manifestations of terrorism.[57] Indeed, 'terrorism' in the Philippines must be examined alongside the historical geography of social struggles in the southern part of the country.

Known as the most Islamized region of the Philippines, and also the region with some of the highest levels of poverty (in spite of significant natural resources), the south has been most susceptible to separatist struggles. There has been broad consensus amongst a number of scholars, notably San Juan and McKenna, that current conflicts date back to Spanish and American colonial imperatives in the Philippines.[58] The Spaniards, who were trying to establish theocratic rule over the Southern islands by conquering and converting the indigenous communities, established the bifurcation of the 'Infidels' (Muslim) and the 'Civilized' (Christians). Ethnic difference helped to legitimize the violent exploitation of the natives and the theft of their lands and resources. However, as McKenna points out, the armed Islamic resistance to the Spanish aggression was not a unified one; rather local sultanates often fought with one another, sometimes forging alliances with the Spaniards to do so. When the US annexed the Philippines in 1898, there was continued suppression of the native opposition to attempted colonial rule. Unable to subdue the Muslims through military might, the US negotiated tactical compromises with the local leaders, coaxing their support through various concessions. Ironically, it was however the American colonial authorities that encouraged the development of a transcendental ethno-religious identity among Philippine Muslims. This unified identity then formed the basis of the nationalist 'Bangsamro' identity of the Muslim separatist movement in the late 1960s. The

56 Butler (note 5).

57 Said (note 2); Gregory (note 4).

58 T. McKenna, *Muslim Rulers and Rebels* (Berkeley, CA: University of California Press 1998); E. San Juan, 'Ethnic Identity and Popular Sovereignty: Notes on the Moro Struggle in the Philippines', *Ethnicities* 6/3 (2006) 391–422.

underlying intention seems to have been to prepare Philippines Muslims for the eventual end of American colonialism and their inclusion in an independent Philippine republic as a consolidated and relatively progressive minority. It was a naïve intention but such colonial practices did have the effect of encouraging the development of a unified Philippine Muslim identity. Not incidentally, American colonialism also provided the *lingua franca* – English – for contemporary Muslim separatists.

In addition, postcolonial politics in the Philippines also exacerbated contributions to the Muslim separatist movement. Soon after the founding of the Republic in 1946, the Philippine government began to encourage large scale migration from the poor regions of the north to the agricultural frontiers of the lightly populated southern islands. The large, fertile island of Mindanao became the primary destination for Christian migration to Southern Philippines and by the late 1960s, Mindanao Muslims found themselves a relatively impoverished minority in their own homeland. While the scale of Christian immigration to Mindanao itself caused inevitable dislocations, the manner of its occurrence also produced glaring disparities between Christian settlers and Muslim farmers. This is due to the fact that government services available to the Muslims were not only meager (as compared to those obtained by immigrant Christians) but were also fewer than they had received under the colonial regime.[59] Thus, as can be seen, these complex layers of historical events have resulted in the politico-economic dispossession and cultural discrimination of the Muslim peoples in Southern Philippines. It was under such conditions that the Moro National Liberation Front (MNLF) came to the fore in 1969. The manifesto of MNLF was to establish a revolutionary Bangsamoro Republic whose goal is complete independence – that is freedom from the

> terror, oppression and tyranny of Filipino colonialism which has caused us untold sufferings and miseries by criminally usurping our land, by threatening Islam through wholesale destruction and desecration of its places of worship and its Holy Book and murdering our innocent brothers, sisters and folks in a genocidal campaign.[60]

However, the ideal of full sovereignty for the Bangsamoro nation, via the mediation of the Organization of Islamic Conference (OIC) which conferred legitimacy on the MNLF, has devolved into the reality of an 'autonomous unit'. After the failure of the 1976 Tripoli Agreement, the MNLF have settled for autonomy instead of genuine independence which was the original objective. This led to the signing of a Peace Agreement with the Ramos government in 1996 which arranged for the implementation of the original Tripoli Agreement in two phases. First it created a transitional administrative structure known as the Southern Philippines Council

---

59   McKenna (note 60).
60   San Juan (note 60).

for Peace and Development. The second phase, which was scheduled to begin in 1999 called for the establishment of a new regional autonomous government. However, the implementation of this agreement has been stalled badly and is in serious danger of unraveling altogether. It is still stuck in the initial phase and has made very little progress in achieving peace or development for the Muslim South.

This compromising stance of MNLF did not appease many of its members who eventually decided to leave the organization to pursue an alternative vision. MILF was thus formed to strive for a separate Islamic state in Southern Philippines, demanding political, economic and military self-reliance. Having demonstrated its capability to wage interminable warfare, the MILF has signed a truce with the Arroyo Administration and is now negotiating for land redistribution, recognition of Shari'a law, rehabilitation of war-ravaged areas, implementation of previous agreements begun by the MNLF government and so on. However, the progress of the peace talks has been slow and doubts have been raised about the sincerity of the Philippine government in recognizing the legitimacy and needs of the Muslim peoples. Besides, the failure of the Peace Agreement between MNLF and the Philippine government has also led to serious misgivings about the futility of such treaties. Such frustrations, and the suspension of the secessionist agenda by both MNLF and MILF, may push more Muslims to the ASG, given that it can now claim to be the sole representative in realizing the dream of an independent Islamic state for the Bangsamoro nation. As San Juan puts it, despite its terrorist and extremist sectarian means, the ASG with its 'sloganeering can engage actors and agencies in the discursive field of ideological contestation so long as the desire for emancipation from Christian chauvinism – rampant if hidden throughout society – and full enjoyment of its life-form defined by Islamic values and rituals cannot be satisfied by existing political order'.[61]

From the Philippines case, it can be seen that a single overarching global network with regional franchises does not drive contemporary 'terrorism' in the country; rather the complex connectivities of current violence/conflicts to the dynamics of geo-historical specificities cannot be under-rated. Effective responses thus need to take into account how those who understand themselves as victims experience these geo-historical conditions and how they enter into their own formation as acting and deliberating subjects. While such analyses are beyond the scope of this chapter, they can no doubt form the basis for alternative accounts that provide more nuanced portrayals of the contours of violence in South East Asia.

Conclusion

This chapter has critically assessed the meanings of and responses to the demarcation of South East Asia as the 'second front' in the global war on terror. I have attempted

---

61   Ibid., p. 405.

to demonstrate that the 'second front' rhetoric is counterproductive for rooting out the sources of terror in the region. It is a meta-narrative that not only distorts our understandings of violence in South East Asia but also our responses to it. Indeed by extending Gregory's conceptual triad of locating, opposing and casting out to the South East Asia setting, I have elucidated how such imaginative geographies direct performances to a single strategic approach, focusing on training and other centralized strategies. The capacities of these, however, to respond in a relevant manner are seriously constrained by the diversity of sequences and motives for violence. In lieu of such inadequacies, I have followed Gregory in a plea for the development of counter-geographies that seek to provide alternative (and less destructive) maps of meanings. By *contextualizing* my analysis in the Philippines, the diverse webs of *connection* and *affiliation* that tie the present moment of conflict and violence in the country to its multiple histories and geographies are illuminated. The excavation of these contrapuntal geographies provides an opportunity for a genuine understanding of the underlying *conditions* that give rise to 'terrorism' and get to the 'root' of violence.

Towards the end of *The Colonial Present*, Gregory engages in pensive philosophical and ontological reflections on the possibility of a more humane and peaceful world. While he acknowledges that there will still be disagreements, conflicts and enemies, he calls for a geographical imagination that can 'enlarge and enhance our sense of the world and enable us to situate ourselves within it with care, concern and humility'.[62] Urging the formation of cosmopolitan geographies, he argues that there is a need to dislocate from First World privilege so as to even begin to imagine a world in which that violence might be minimized, in which an inevitable *interdependency* becomes acknowledged as the basis for global political community. It is such alternative imaginings of the future that we need to uphold and strive towards instead of allowing the perpetuation of violence through problematic names (such as the 'second front', war on terror) that restrain us from thinking and acting responsibly in our everyday lives.

Acknowledgements

Thanks are due to Jenny Kynaston who drew the figures for this chapter.

---

62   Gregory (note 4) p. 262.

PART 2
Governing Through Security

Chapter 6
Disciplining the Diaspora:
Tamil Self-Determination and
the Politics of Proscription

Suthaharan Nadarajah

Introduction

The phenomena – or, more precisely, the discourse – of 'terrorism' has increasingly
come to dominate contemporary international politics. From the alliance building
of states waging a 'global war on terror' to their domestic measures combating
'terrorism financing' or even curbing the 'glorification of terrorism',[1] a myriad of
terrorism-related techniques, technologies and strategies have evolved in recent
times. Whilst these trends have inevitably intensified since the deadly attacks
on the United States on 11 September 2001, they have long been emerging,
particularly since the collapse of the Soviet Union, after which global liberalism
became a strategic project pursued by powerful Western states and their associated
institutions and organizations.[2] Much disputed but now quite unavoidable, the
concept of 'terrorism' has thus led to a plethora of new power relations which
redefine the fields of possibilities for a range of actors around the world, including
states, NGOs, business, media, political entities, individuals, both citizen or
immigrant, and, of course, armed non-state actors. These ever expanding and diffuse
relations of power are producing new subjectivities, new forms of domination and,
inevitably, new forms of resistance.

This chapter is an exploration of this phenomenon as it impacts on the question
of Tamil self-determination as, played out in relationships between the Tamil
diaspora and Western host states. In particular, it examines the consequences
of the designation of the Liberation Tigers of Tamil Elam (LTTE), an armed

---

1 *The Guardian*, 'MPs Back Ban on "Glorification" of Terrorism' (15 February 2006)
at <http://www.guardian.co.uk/terrorism/story/0,,1710278,00.html>.

2 T. Young, '"A Project to be Realised": Global Liberalism and a New World Order',
in E. Hovden and E. Keene (eds), *The Globalization of Liberalism* (Houndmills: Palgrave
2002) 173–93. See also M. Dillon and J. Reid, 'Global Governance, Liberal Peace and
Complex Emergency', *Alternatives* 25/1 (2000) 117–43 as well as their 'Global Liberal
Governance: Biopolitics, Security and War', *Millennium: Journal of International Studies*
30/1 (2001) 41–66.

opposition movement that many Tamils consider to be an authentic expression of resistance to oppression by the Sinhala-dominated Sri Lankan state, as a terrorist organization, a process that has intensified since the beginning of the war on terror. The implications of this designation and the conflicting ideas of legitimacy and resistance that it calls into play are well exemplified by Plate 6.1, which shows opposing demonstrations by expatriate Tamils and Sinhalese that took place on 10 June 2008 outside the Commonwealth Secretariat in London while Sri Lankan President Mahinda Rajapaksa was meeting other leaders. The Tamils, protesting against '60 years of oppression' demonstrated in support of self determination while the members of the Sinhala community challenged the legitimacy of this demand. While the nature of relationships between Western states, the Tamil diaspora and the politics of proscription is considerably more complex than is suggested by this picture, it captures some of the contradictions in the British state's approach to Sri Lanka's conflict: whilst the UK deems the LTTE a terrorist organization using violence against a multicultural democracy, British police erect barricades to separate the polarized Tamil and Sinhala communities.

Plate 6.1 Opposing demonstrations by expatriate Tamils and Sinhalese
outside the Commonwealth Secretariat in London, 10 June 2008
Source: Photograph by Suthaharan Nadarajah.

The concept of terrorism has impacted in innumerable ways on the lives of large numbers of ordinary people around the world and, especially, on the activities, strategies and identities of countless entities engaged in politics, broadly defined. This is despite manifest disagreement on what actually constitutes 'terrorism'. The determined efforts of numerous scholars, research centres and institutions, not least the United Nations, are yet to produce a consensus and even the impatient insistence of leading states has not impelled progress towards a universally acceptable definition. Yet, it is precisely this undefined, even indefinable, trait of 'terrorism' – along with its acutely menacing characteristic – that underpins the productive and subjectifying capacity of numerous apparatuses of domestic and international governance. To begin with, as Jenny Hocking puts it,

> through its carriage of an implicit yet overwhelming moral illegitimacy, the language [of terrorism] itself neatly averts consideration of complex questions of causation by assigning an apparently uncontested meaning to diverse incidents of political violence. It is a powerful terminology and one which allows for the ready adoption of extreme measures that would otherwise be strongly resisted.[3]

However, despite the abhorrence and fear it knowingly invokes[4], the discourse of terrorism 'constructs no precise criminal act, rendering terrorism of dubious utility as a legal entity'.[5] Notwithstanding this, apart from their condemnatory rhetoric, it is primarily through specialist – and usually draconian – legislation that Western states now *constitute* acts of violence as terrorism. In other words, apart from the vocabulary of 'terrorism'[6] it is states' use of anti-terrorism legislation to confront specific acts of violence that serves to *define* these acts as 'terrorism.' If a building is torched, the difference between an act of arson and an act of terrorism is the legislation under which the perpetrators are charged – more so, even, than the rationales put forward by those responsible.

Similarly, armed organizations now become defined as terrorists not just when condemned as such by powerful states, but when they are *proscribed* under anti-terrorism legislation. This is especially so when the violence of the organization in question is not directed at the proscribing states or even conducted within their territories. Thus, although proscription is effected through the framework of law, it is a patently *political* act, one moreover which decisively moves the organization *and* further debates around its stated goals out of the space of politics into that

---

3   J. Hocking, 'Counter-Terrorism and the Criminalisation of Politics: Australia's New Security Powers of Detention, Proscription and Control', *Australian Journal of Politics and History* 49/3 (2002) 355–71, p. 359.

4   R. Jackson, *Writing the War on Terrorism: Language Politics and Counter-Terrorism* (Manchester: Manchester University Press 2005).

5   Hocking (note 3).

6   Richard Jackson has referred to this as the 'rhetorical construction of reality' (note 4) p. 2.

of crime and security. Whilst the proscribing state does not have to establish the *legal* case for the ban prior to its enacting, *deproscription* can only be achieved by a successful legal challenge: the burden of proof rests solely with the accused. Notably, quite apart from its acts of violence, proscription delegitimizes the organization itself and, therefore, its salience to the political arena in which it operates. As Hocking's observation above notes, the language of terrorism effaces the context, such as resistance to state oppression or racial persecution, in which non-state violence often takes place. Proscription, moreover, also forecloses or deters further discussion of this effect – outside a legal challenge, that is.

The focus of this chapter is, however, not on armed movements themselves, but on how the international 'anti-terrorism' regime that has emerged over the past two decades enmeshes with other domestic apparatuses and discourses in Western liberal states to actively shape the behaviour of a range of other, *unarmed* political actors. In particular, it considers how the discourse of terrorism serves to 'conduct the conduct'[7] of actors who are politically active within Western liberal democracies, for example in terms of advocacy in connection with foreign conflicts, and, by extension, the behaviour of residents, citizen or otherwise, who hail originally from these distant places.[8] Using the Tamil diaspora as a case study,[9] the chapter examines how the international anti-terrorism regime, gradually extended since the late nineties to include the LTTE,[10] redefines the field of possibilities for ordinary Tamils vis-à-vis their struggle against state oppression in Sri Lanka. Although each country's proscription has different legal implications and has been imposed for different stated reasons, individually and collectively they serve to criminalize the LTTE *and its political project*, especially amid the explicit criticism of both by the proscribing states. Inevitably this has raised serious implications, both within host countries and in Sri Lanka, for the activities, political or otherwise, of other Tamil organizations and individuals. The central claim of this chapter is that the anti-terrorism regime both directly, by coercively shutting down some political spaces, and indirectly, by providing alternative 'safe' spaces, shapes Tamil expatriates' political activity in their *hostlands* towards

---

7   M. Foucault, 'Governmentality', in G. Burchell, C. Gordon and P. Miller (eds), *The Foucault Effect: Studies in Governmentality* (Hemel Hempstead: Harvester Wheatsheaf 1991) pp. 87–104.

8   Whilst not taken up further here, it is worth noting how the dynamics of terrorism-related suspicion and regulation apply to specific populations within Western states such as Tamils, Kurds, Muslims, etc., rather than the wider host publics.

9   The chapter is based on interviews conducted as part of research for the author's doctoral thesis between 2003 and 2007 with diaspora activists and representatives of Tamil community organizations in Australia, Canada and Europe, as well as officials from Western government departments.

10   Terrorism-related bans on the LTTE under domestic law include the US (1997), UK (2001), Canada (2006) and the EU (2006). India banned the LTTE in 1991. Interestingly, since 2002 the LTTE has not been banned in Sri Lanka; Colombo outlawed the LTTE in 1979 and again in 1998, lifting the ban before peace talks in 2002.

realizing a specific – liberal governmental – vision for *Sri Lanka*.[11] Moreover, it is the inherent ambiguity of the notion of 'support' for 'terrorism' that allows a raft of governmental techniques and technologies, operating alongside the disciplinary framework of anti-terrorism, their purchase on target populations within Western liberal states. It is worth noting, moreover, that when enmeshed with ruthlessly restrictive asylum and immigration regimes, the anti-terrorism regime also shapes the political and other activities of Tamils beyond the hoststate's borders: 'supporting' a banned organization is now sufficient grounds for exclusion.

As a minimum, the behaviours being compelled from Tamil expatriates include rejection of armed struggle and taking up of political struggle in Sri Lanka through non-violent modalities – no matter how ineffectual these might actually be in ending state repression. Going further, the international anti-terrorism regime also induces other behavioural changes amongst politically active Tamils, including refocusing their efforts towards the pursuit of 'good governance' rather than 'self-determination', a concomitant accordance of primacy to 'human' rather than collective rights and so on. Core issues confronting Tamils in Sri Lanka – such as institutionalized ethnic discrimination, embedding of Sinhala majoritarianism within the state bureaucracy and military, absence of physical security and protection of law, lack of media freedom, and so on – thus shift from being justifying elements of their demand for *independence* to being targets for *reform* of the Sri Lankan state.[12] In other words, rather than constituting racist state persecution to be escaped through independent statehood, these elements become reconstituted as failures of governance that must be corrected through state reform. In short, the international anti-terrorism regime contributes through diffuse, peripheral channels towards the transformation of the Tamil liberation struggle into its very antithesis: the *strengthening*, through the logic of 'reform', of the hitherto rejected Sri Lankan state. This is not to say this radical change has been achieved; indeed this aspect of the global liberal project is far from complete and recent developments, including deepening state repression in Sri Lanka and the post-proscription evolution of diaspora strategies, have contributed to an intensification of Tamil demands for independence: indeed the increasing use of anti-terrorism measures, including force, can be seen as a shift from governmental to disciplinary efforts and the

11   Foucault (note 7). For a detailed discussion, see M. Dean, *Governmentality: Power and Rule in Modern Society* (London: Sage 1999).

12   These have been discussed in the extensive literature on Sri Lanka. See for example S. Bose, *States, Nations, Sovereignty: Sri Lanka, India and the Tamil Eelam Movement* (London: Sage 1994); S. Krishna, *Postcolonial Insecurities: India, Sri Lanka and the Question of Nationhood* (Minneapolis: University of Minnesota Press 1999); R. Herring , 'Making Ethnic Conflict: the Civil War in Sri Lanka', in M. Esman and R.Herring (eds), *Carrots, Sticks and Ethnic Conflict: Rethinking Development Assistance* (Ann Arbor: University of Michigan Press 2001) 140–74; J. Goodhand, *Aid, Conflict and Peace Building in Sri Lanka* (London: The Conflict, Security and Development Group 2001); V. Nithiyanandam, 'Ethnic Politics and Third World Development: Some Lessons from Sri Lanka's Experience', *Third World Quarterly* 21/2 (2000) 283–311.

exercise of sovereign power. However, an examination of the microphysics of the securitization of politics that the terrorism discourse entails serves to illustrate the wider transformative potency of Western states proscribing a foreign armed organization as terrorists.

Shaping, Not Curtailing Politics

To begin with, contemporary (typically Western-led) international efforts to end intra-state conflicts take place in the framework of establishing what Mark Duffield has labelled 'liberal peace'.[13] These endeavours, increasingly undertaken as part of the wider global liberal project by leading Western states and their associated organizations, institutions and agencies, posits economic interdependence, democracy and the rule of law as constituting the sustainable foundations for world peace.[14] As such, liberal peace is 'irrevocably linked to the territorially sovereign state as an umbrella for political community'[15] and therefore to its stabilization and strengthening against challenges to its authority, an imperative reflected in contemporary anxieties over 'failed' or 'fragile' states (see Chapter 3).[16] Thus, even in the context of minority demands for independence from repressive majoritarian states, what is deemed desirable is not the 'fragmentation' of the state along 'ethnic lines' but rather its strengthening and reform in liberal governmental terms, that is towards a single democratic state with strong liberal institutions, a civic polity and an open economy. (Even in the exceptional instances where new states have emerged, such as in the former Yugoslavia, this has happened through probationary periods of international trusteeship in which, as rites of passage, these tenets of liberal peace have to be adopted and ingrained.)[17]

By the same logic, armed movements, including those fighting for national liberation or self-determination, are considered, first and foremost, *threats* to the state that must be compelled to disarm and seek remedy for the grievances they claim to represent through non-violent means.[18] Thus, quite apart from confronting anti-*Western* violence, it is in the worldwide pursuit of liberal peace that the 'War on Terror' has became a 'global' struggle, drawing disparate conflicts in different locations into a single conceptual framework. It is also in this

---

13   M. Duffield, *Global Governance and the New Wars* (London: Zed Books 2002). For a detailed discussion of the concept also see O. Richmond, *The Transformation of Peace* (Houndmills: Palgrave Macmillan 2007).

14   S. Willett, 'New Barbarians at the Gate: Losing the Liberal Peace in Africa', *Review of Political Economy* 106 (2005) 569–94.

15   Richmond (note 13) p. 13.

16   M. Duffield, *Development, Security and Unending War* (Cambridge: Polity 2007).

17   Richmond (note 13); M. Pugh, 'The Political Economy of Peacebuilding: A Critical Theory Perspective', *International Journal of Peace Studies* 10/2 (2005) 23–42.

18   Richmond (note 13) pp. 81–3.

context that 'terrorism' is no longer a label the Western liberal democracy applies only to domestic threats or foreign attackers. In the past such states have been territorially conservative when outlawing armed non-state actors, with violence in foreign sites not a direct concern except when their own interests and nationals, or perhaps those of close allies, were being targeted. The political wings of armed organizations active in other parts of the world could therefore function freely alongside other dissidents in the territories of Western democracies, often to the chagrin of the states they were opposing. However, this has changed markedly in recent times. Western states, as a matter of routine, now not only condemn non-state violence in far away places as 'terrorism', but also actively respond to it at home with crackdowns against the groups held responsible and their supporters. The shift is exemplified by a British judge's 2007 observations while upholding the conviction under his country's terrorism laws of a Libyan dissident accused of supporting violence against the Ghaddafi regime: 'We can see no reason why', he asserted, 'the citizens of Libya should not be protected from such activities by those resident in this country in the same way as the inhabitants of Belgium or the Netherlands or the Republic of Ireland'.[19] That the regimes being confronted by such 'terrorism' are also sometimes characterized by the proscribing states as repressive or non-democratic is thus irrelevant. So, for that matter, is whether the violence is directed at military or civilian targets.[20] Thus it is not simply a question of Cold War-style solidarity between allied states, but the blanket hostility to non-state violence inherent to global liberalism.[21]

Despite acknowledging that the LTTE's armed struggle for independence is waged solely against the Sri Lankan state, most Western states have now proscribed the movement as a terrorist group and included it in their domestic anti-terror actions.[22] This has taken place, notably, amid a growing recognition that the global Tamil diaspora, numbering over 800,000,[23] is a key factor in the Sri Lankan conflict. Tamil expatriates are vocal advocates of self-determination and independence, providers of humanitarian relief for the war- and tsunami-affected

---

19   R v. F [2007] EWCA Crim 243 at <http://www.bailii.org/ew/cases/EWCA/Crim/2007/243.html>.

20   British Home Secretary's response, citing section 3(5)9a of the Terrorism 2000 Act, to the British Tamil Forum's appeal against the proscription of the LTTE (11 January 2008).

21   Richmond (note 13) pp. 81–3. See also A. Guelke, *The Age of Terrorism and the International Political System* (London: I.B. Tauris 1998).

22   See note 10.

23   Whilst there are no firm recent figures for Tamil Diaspora numbers, in June 2001, the UNHCR estimated the stock of internationally displaced Tamils to be 817,000, most of whom are/were refugees or asylum seekers. Canada topped the list, hosting an estimated 400,000 Tamils, followed by Europe (200,000), India (67,000), the United States (40,000), Australia (30,000), and another 80,000 living in a dozen other countries. Figures cited in D. Sriskandarajah, 'The Migration–Development Nexus: Sri Lanka Case Study', *International Migration* 40/5 (2002) 283–307.

Tamil areas of the island and, in particular, are a source of financial, moral and political support for the LTTE.[24] The diaspora has thus increasingly become a key target and also a *vehicle* for global liberal governmental efforts. Although the language of terrorism has long been used by international actors in relation to the LTTE,[25] now, more so than at any time before, the discourse of terrorism has come to mediate the multifaceted relationships between expatriate Tamils, their host states and populations, the LTTE and other international actors. To examine how the international anti-terrorism regime, encapsulating or permeating a range of domestic governmental apparatuses, actively shapes Tamils' conduct, this chapter considers what might otherwise be an unremarkable area of activity in the West: advocacy at home for political causes abroad.

In keeping with its oft-reiterated 'with us or against us' rhetoric, the international anti-terrorism discourse places political actors – not just armed organizations – into binary categories of acceptable and unacceptable. There are gradations within such distributive categories,[26] but the legal act of proscription defines a clear step of exclusion: just as being denounced as terrorists is one thing and being banned quite another, being criticized for specific 'extreme' views is quite different to being prosecuted for 'supporting' or 'glorifying' terrorism. The difference, crucially, depends more on the subjective opinion of the authorities than specified or self-evident criteria. It is not simply a question of not being able to express support for outlawed violence, say by invoking the principles of self-defence against genocide or resistance to state repression. Rather, proscription comes to have a much wider impact in the 'global' political space that banned armed organizations function within by enabling the categorizing of *political* positions or goals as acceptable and unacceptable (using the labels of 'moderate' and 'extremist' for example). There is not necessarily an automatic mapping: bans of the 'reprehensible' LTTE have sometimes been accompanied by assertions by the proscribing states that the organization has 'legitimate goals'[27] (that is, Tamil 'grievances'). However, the international discourse around Sri Lanka has also long held the demand for Tamil independent statehood to be 'extreme' and that for federalism or autonomy (i.e. accepting Sri Lankan sovereignty) to be a 'moderate' position. With the Tamils' claim of the right to self-determination (that is, that the appropriate form of governance for them is *their* prerogative) remaining unresolved, such international assertions are arbitrary acts of power – that is, of politics – rather than findings of

---

24   D. Byman, P. Chalk, B. Hoffman, W. Rosenau and D. Brannan, *Trends in Outside Support for Insurgent Movements* (RAND 2001) at <http://www.rand.org/pubs/monograph_reports/2007/MR1405.pdf>.

25   S. Nadarajah and D. Sriskandarajah ,'Liberation Struggle or Terrorism? The Politics of Naming the LTTE', *Third World Quarterly* 26/1 (2005) 87–100.

26   M. Brigg, 'Post-Development, Foucault and the Colonisation Metaphor', *Third World Quarterly* 23/3 (2002) 421–36, p. 428.

27   *Tamil Guardian*, 'US: LTTE Has "Legitimate Goals, Unacceptable Methods" Report Compiled Using Agency Reports' (14 June 2006).

international law. Nonetheless, in the context of organizations like the LTTE being proscribed, these *assertions*, rather than international law, come to emphasize the acceptability or otherwise of political stances taken up within Western liberal states. For example, an oft-stated assertion by the United States, the first Western state to outlaw the LTTE, is that to be deproscribed, the organization must first renounce violence 'in word and deed' *and also* give up its goal of an independent Tamil Eelam. Addressing a 2002 donor conference at which an LTTE delegation was participating as part of the Norwegian led peace process, US Deputy Secretary of State Richard Armitage stated:

> Let me leave no doubt: my nation stands firm in the resolve that the tactics of terror can never achieve legitimate aspirations. ... We urge the LTTE ... to make it clear to the people of Sri Lanka and indeed to the international community that the LTTE has abandoned its armed struggle for a separate state; *and instead accepts the sovereignty* of a Sri Lankan government that respects and protects the rights of all its people.[28]

It is this *conflation* of armed actors' violence ('terrorism') with specific political goals which global liberalism finds unacceptable that enables the categorization and sorting of other actors who, whilst not involved in armed struggle, are active in the same political space: to demand Tamil Eelam is to be deemed sympathetic to the LTTE. Crucially, it is this conflation, more than anything else, which enables 'supporters of terrorism' to be identified and labelled as such. This is not to say that taking up specific political stances will directly result in prosecution under anti-terrorism legislation. Rather, it is to suggest that taking up such positions is to *risk* drawing the invasive attention of terrifying and extensive state security apparatuses, as well as exclusion and marginalization from legitimate domestic spaces, such as lobbying access to centres of power. This 'categorical suspicion'[29] also paves the way for officialdom's resistance to, if not obstruction of, specific instances of legitimate activity such as staging rallies and public meetings, leafleting, public broadcasting, and so on. Inevitably, it especially raises serious difficulties for fund raising in support of such political activities. At an individual level, it can raise difficulties in travel (e.g. visas and work permits, 'no-fly' lists, etc), employment (certainly in the military and defence industry, but also in civil service or private industry roles where even low levels of security clearance are needed) and so on.

Thus, it is this implied equivalence of unarmed actors' *political* positions to support for the *violence* of 'like-minded' armed actors that allows a variety of

---

28   Address to Aid Conference in Oslo, Norway (25 November 2002) quoted in *TamilNet*, 'US Calls to Renounce Violence' (emphasis added) at <http://www.tamilnet. com/art.html?catid=13&artid=7892>.

29   G. Marx, *Undercover: Police Surveillance in America* (Berkeley: University of California Press 1988).

governmental techniques and technologies within Western states their purchase. Moreover, it is amid this implied equivalence between specific political positions and support for 'terrorist' violence that coercive apparatuses enabled by the discourse of (anti-)terrorism come to induce positional and behavioural *compulsions* amongst political actors in the West. In the wake of the proscription of an armed organization, the central question other unarmed actors come to face is how to engage in politics without incurring the ostracizing and manifestly dangerous label of 'supporters of terrorism'. At a basic level, if they are to safely pursue their political goals, they must adopt specific behaviours including discernibly distancing themselves from, even condemning, the 'terrorists' fighting for these goals. As in the case of the Tamils, the implications for a people remaining united while seeking national liberation from racial oppression are obvious. Moreover, by enacting legislation that threatens severe punishment for 'supporting' terrorism, the terms of what does and doesn't constitute acceptable *advocacy* are also set. For example, if a proscribed organization enters into negotiations with its state adversary, how can its stances on the issues under negotiation be endorsed by other Tamil actors – especially if the proscribing state strongly opposes these positions? If the organization breaks off peace talks and resumes its military campaign citing reasons (say non-implementation of agreements already reached or continuing state violence) that host states deem unacceptable, what are the consequences of echoing these reasons? As noted above, it is not a question of whether a particular viewpoint has merit or not in itself, but whether it is possible to articulate it without fear of being punished for supporting the *violence* (that is, terrorism) of armed actors also holding these views. The range of advocatable positions available to unarmed political actors thus narrows and is sometimes closed off altogether by the outlawing of other *armed* actors.

Crucially, the anti-terrorism regime's coercive effects not only close off some spaces, but also *promote* other, more preferable, political positions. Just as it discourages certain behaviours, such as advocacy of anti-state violence and the championing of 'extreme' positions (such as independence), it also *encourages* the taking up of other, more 'moderate' positions. Thus it is not simply a question of silencing or curtailing debate, but the more productive effect of shaping and directing its terms. With its punitive apparatuses poised over some spaces and withheld from others, the international anti-terrorism regime promotes the articulation and adoption by the target population of a range of *liberal governmental* positions. At a basic level these include, as noted above, rejection of armed struggle and the adoption of non-violent methods. However, as in the case of Tamil expatriates, the regime goes much further, for example encouraging the politicization of what are deemed 'inclusive' rather than 'exclusive' identities – for example adoption of a 'civic' (Sri Lankan), as opposed to 'ethnic' (Tamil) one, accordance of primacy to human, rather than collective, rights and so on. In short, by setting out what are 'moderate' and 'extreme' views and simultaneously wielding the punitive apparatuses enabled by proscription, the international anti-terrorism regime constitutes the disciplinary and sovereign framework required for the expansion of liberal governmentality.

Disciplining the Diaspora

This illiberal aspect of the anti-terrorism regime is no longer disguised, as exemplified by the harsh penalties dealt out in Britain for those held to be 'encouraging' or 'glorifying' terrorism.[30] This might, at a first glance, seem a contradiction to the notion that the regime seeks to promote liberal values abroad. However this inconsistency fades when considered in the context of the global liberal project which posits these 'universal' values as an unalloyed (and thus incontestable) ultimate good, in the interests of which extreme, patently illiberal measures, including massive military violence, are justified.[31] Terrorism proscriptions enable a domestic disciplinary framework, one based on apparatuses of coercion, surveillance and, as discussed below, behavioural training, that seeks to actively produce well behaved citizens of liberal governmentality. As proscription is primarily a legal sanction, the coercion is underwritten by poised state machinery: police and other internal security forces, prisons, deportations, individual sanctions (including restrictions on travel, employment, access to welfare, etc.) and so on. Beyond these, the much publicized practices of 'rendition' and other extra-judicial aspects of the 'global' anti-terrorism architecture (including the sharing of 'terrorism intelligence' with other states) are amongst the fearsome consequences of crossing the indiscernible and ever shifting line of 'supporting terrorism' whilst living in the West. Thus the disciplinary framework is closely linked not just to governmentality but also to sovereign power.

Terrorism-related surveillance begins with the overt deployment of state apparatus, including the police, intelligence services and other regulatory structures. Public rallies and demonstrations often require police approval, which is frequently not forthcoming for events organized by pro-Eelam Tamil actors but is easier for 'moderate' actors to obtain. Even when approved, events are increasingly subjected to overt monitoring, with grim-faced uniformed police photographing and videotaping those attending as well as those on stage. In the context of manifest Western, opposition to the Tamil Eelam project, such menacing security presence inevitably induces a high degree of self-regulation amongst both speakers and attendees, as well as a reluctance to participate in some events. Slogans and placards must be careful to avoid being *seen* as supporting (or in the UK, 'glorifying') terrorism. Speeches must also remain within these unspecified limits, which, as noted above, are more subjective assessments by authorities than

---

30   See, for example, Amnesty International's criticism of drafts of the Terrorism Act (2005) at <http://www.amnesty.org/en/library/asset/EUR45/038/2005/en/dom-EUR45 0382005en.html>.

31   Dillon and Reid (note 2).

clearly defined parameters. The text of leaflets, publications and even Internet websites must be 'self'-regulated in the same manner, as must reporting and commentary by Tamil expatriate media.[32]

In addition to this formal oversight is the state's surveillance 'at a distance' i.e. its efforts to co-opt the citizenry as part of its terrorism-related scrutiny. Public advertisements – posters, radio broadcasts, etc exhorting citizens to 'report anything suspicious' (and encouraging them to err on the side of caution) heighten the gap between those clearly unproblematic citizens and those who *might* be supporters of terrorism. For example, an advertisement by the British police in a London newspaper urged the city's residents: 'Terrorism – if you suspect it, report it.'[33] Listing the number of a 'confidential, anti-terrorist hotline', it goes on to list what might be suspect, including: 'Terrorists use computers; do you know someone who visits terrorist-related websites? Terrorists need to travel; Meetings can take place anywhere. Do you know someone who travels but is vague about where they are going? Terrorists need transport: if you work in vehicle hire or sales, has a sale or rental made you suspicious?'[34]

Thus, whilst proscription might be a legal matter, accusations of supporting or assisting terrorists can easily be levelled in fora *other than* the courts. Indeed, anonymous accusations by anyone (and anywhere) are positively encouraged. The government of Sri Lanka or one of its departments, such as the local embassy, hostile media (local or foreign, including Sri Lankan), political, business or personal rivals, and otherwise disgruntled individuals, etc. can level accusations, no matter how unsubstantiated, against Tamil organizations and individuals. Such dynamics and their consequences have been explored in the extensive literature that comprises surveillance studies[35] and in studies of welfare regimes.[36] Notably, there is usually little or no risk of penalty to false accusers. It is often impossible to trace the pointing finger that triggers invasive probes into political activities and, indeed, personal lives. Even if accusers are identified, given the opaqueness of the reporting mechanisms, there is no redress save that offered by difficult and probably expensive libel claims. The proscribing state may choose not to or may not have the resources to monitor every broadcast or article by every media outlet, scrutinize every publication by every organization, or listen to every speech. But

---

32   As mentioned in note 8, Tamil media experience different constraints to mainstream Western media. For example, although UK legislation even forbids publication of the logo of a banned terrorist organization, the LTTE logo appears on the BBC website's reports, for example.

33   Metropolitan Police/British Transport Police advertisement in *London Lite* (23 March 2007) p. 8.

34   Ibid.

35   See for example, D. Bigo, 'Security and Immigration: Toward a Critique of the Governmentality of Unease', *Alternatives* 27 (2002) 63–92.

36   See for example, B. Cruikshank, *The Will to Empower* (New York: Cornell University Press 1999).

anyone who suspects or claims a breach of anti-terrorist legislation can alert the authorities, even anonymously, without fear of repercussion. Expatriate Tamil electronic media, for example, routinely have to cooperate with regulators following up spurious claims filed anonymously that their broadcasts have supported the LTTE or its violence.[37]

That this 'surveillance at a distance' is an integral part of the proscription regime was underlined by Canada's government in April 2006. Announcing his country's ban on the LTTE, Public Safety Minister Stockwell Day also unveiled an advertising campaign directed at local Tamils to 'explain' the terrorist designation: 'in one of the ads that we're putting out there's a number listed for people who are *emboldened now* by the [ban] and may want to contact authorities for follow up'.[38] He added: 'authorities will be keeping a close eye on *affiliated* organizations that *could* run afoul of the terrorist designation should they continue *allying* themselves with the [LTTE]'.[39] The immediate question for Tamil political actors is what constitutes the threatening characterization of an 'affiliated organization'. More importantly, in the context of taking up political positions in Canada, is what exactly 'allying' themselves with the LTTE might mean – although the obvious inference, of course, being support for Tamil Eelam.

At other times, peaceful Tamil political activity characterized in this way has fallen more directly within the ambit of the legal system. Two of the organizers of a Tamil expatriate rally in London's Hyde Park on 25 July 2006 marking the anniversary of the July 1983 anti-Tamil pogrom in Sri Lanka (an event that was attended by some 15,000 people – see Plate 6.2) were subsequently arrested (exactly a year later) and charged under the Terrorism Act 2006 with organizing an event in support of a banned terrorist organization. Whilst the organizers had been granted permission for the event and police were also in attendance, the investigation into 'supporting terrorism' was initiated by complaints from the Sri Lankan High Commission in London.[40]

---

37   Interviews with staff at the London-based IBC radio and other outlets. Also see British regulator Ofcom's comments on investigations into allegations against music broadcast on Sangamam (ETBC) radio. Bulletin Issue 47 (11 November 2005) at <http://www.ofcom.org.uk/tv/obb/prog_cb/obb_47/>.

38   *National Post*, 'Canada Adds Tamil Tigers to Terrorist List, Urges Tamils to Report on Fund Raising' (10 April 2006) emphasis added.

39   Ibid. Emphasis added.

40   Interviews with organizers, Tamil Youth Organization (TYO–UK) (September 2007).

Plate 6.2 Tamil expatriate rally in Hyde Park, London, 25 July 2006
Source: Photograph by Suthaharan Nadarajah.

Moreover, the ambiguity around what constitutes 'support for terrorism' and individuals' fear of the consequences of falling foul of anti-terrorism laws also paves the way for *state action* outside the courts. Proscriptions of the LTTE have enabled state intervention in Tamil expatriate political and social activity even when no crime is being committed. This is exemplified by similar developments in the Canada, UK, and Europe whereby, after proscriptions of the LTTE (and in Australia, where a fundraising ban was imposed in 2002), security forces have sought to discourage and undermine social and political events organized by pro-Eelam Tamils while leaving unfettered those of actors who either oppose or distance themselves from the LTTE and its political project (it is worth noting here how the lobbying against the LTTE cannot be countered without immediately falling foul of laws outlawing 'support for terrorism'). Police tactics include, for example, privately encouraging, even pressuring, the owners of halls, sports fields and other venues to refuse to hire their sites to pro-independence Tamil organizations.

Such state interventions frequently turn directly on the dissembling logic of the *possibility* of the customer being a 'terrorist front' intending to raise funds through the event. In some cases, venues have admitted to long-standing customers they were now turning away, to having been influenced by suggestions by police that they (venues) *might* become complicit in fund raising for the LTTE. Even when customers point out that their books are audited by the Inland Revenue service and no wrongdoing has been found, many venues are reluctant to take the risk. Indeed, when suggestions of 'supporting' or financing terrorism are raised through such 'unofficial' channels, the event organizers are, inevitably, the least able to reassure venues of their bona fides. The unspecified 'security concerns' that some venues have cited for rejecting Tamil business emerge from the context in which as yet unproven claims of 'links' to terrorism *can* be unproblematically raised – even as a mere possibility – by either the security forces or a range of anti-LTTE campaigners, including the local Sri Lankan embassy, who bombard potential venues with accusations of imminent complicity in 'supporting terrorism'. In some cases, the venues' consequent response has been to urge such problematic customers to bring explicit endorsements from the police – who are, of course, under no obligation to provide these. When approached, police simply advise the event organizers that there are no security concerns about them and that they should appraise the venues thus. Moreover, implicitly and sometimes explicitly, the organizers' *political* positions are acknowledged as the underlying reasons for the difficult situation they find themselves in with the attendant encouragement to retreat from stances that are problematic (see discussion on advocacy below).

At the same time, notably, there is no official bar on any of the *events* themselves – except when permission for rallies or marches are refused, often without explanations having to be offered. Neither are the organizers themselves proscribed or officially blacklisted. By simply raising the spectre of 'supporting terrorism', Tamil organizations can be subject (perhaps even unintentionally on the part of venues) to a subtle array of discriminatory practices. Under other, 'normal', circumstances, venues are under no compulsion to reject customers on the basis of third-party allegations of *impending* criminality; indeed for venues to turn away Tamil or other minority customers on such claims is to risk being sued for racial discrimination. However, in the context of 'supporting terrorism', not only are venue owners able to reject custom, they are *expected* to. More generally, the populations affected by these proscription-related difficulties are invariably minorities, including Tamils, Kurds, Muslims and so on. In short, the anti-terrorism discourse can thus be seen to facilitate the wholesale discarding of the anti-racism and anti-discriminatory safeguards normally applicable to minorities in Western states.

Learning to Speak: Political Advocacy

The impossibility of expressing political support for the LTTE has to be considered in the context of many Tamils seeing the viability of their liberation struggle as linked to the 'success', broadly defined, of the LTTE in resisting and challenging the Sri Lankan state. Having emerged in the eighties as the dominant Tamil protagonist in the conflict following a number of early confrontations within the broader Tamil resistance movement, the LTTE has since developed both a conventional military force and a substantial civil administrative apparatus comprising a *de facto* state in those areas it has established control over.[41] It is the largest and most prominent actor advocating the cause of Tamil self-determination and has been the Sri Lankan government's sole interlocutor in four of the five peace processes (it was part of a coalition in the first round) since the conflict began in the early eighties.[42] However, Tamils in the West cannot explicitly endorse the proscribed LTTE as *representatives* of the Tamil polity in negotiations or other spaces, both for fear of falling foul of domestic law and, amid the 'implicit yet overwhelming moral illegitimacy' of terrorism,[43] to avoid their advocacy efforts being dismissed out of hand as 'extremism' or 'support for terrorism'. It is amid this dynamic of having to avoid political irrelevancy whilst lobbying for the Tamil cause that the *productive* effect of the anti-terrorism regime turns: it is not only a question of self-censorship, but also the specific political positions that come to be *taken up*. This is not to say these constitute deep felt ideological shifts – all too often they do not. Rather, the focus here is how changes in Tamil actors' conduct come to propagate values and positions in keeping with liberal governmentality and write out those linked with national self-determination. There are two separate aspects of this induced shift discussed here: firstly, not positing the LTTE as the Tamil *leadership* that the international community and Sri Lankan state must deal with, and, secondly, reconstituting the specific political terms in which Tamil grievances are framed.

Prior to the proscriptions of the LTTE, a central plank of political activity by pro-independence Tamils in the West has been to endorse and promote the LTTE as the 'sole' or 'authentic' representatives of the Tamil people vis-à-vis resolving Sri Lanka's ethnic question. Especially since the advent of the Norwegian peace process in 2001, Tamil petitions, appeals and messages at mass rallies prior to the proscriptions have reiterated this position, while the Tamil National Alliance (TNA), a coalition of Sri Lanka's four main Tamil parties, put this at the core of its manifesto for the 2004 elections in which it swept the Tamil areas. At the height of the Norwegian-led peace process, in 2003 and 2004, a series of mass rallies – titled 'Pongu Thamil' (Tamil Upsurge) – in every major Tamil population centre in Sri Lanka's Northeast and in diaspora locations endorsed this 'sole representatives' demand. However, the international proscriptions of the LTTE either outlawed further articulation of this stand or where it did not, as in the US (where the right to freedom of speech is, in principle, not trumped by terrorism laws), created a political climate in which it was extremely awkward, if not dangerous, to do

---

41   K. Stokke, 'Building the Tamil Eelam State: Emerging State Institutions and Forms of Governance in LTTE-controlled Areas in Sri Lanka', *Third World Quarterly* 27/6 (2006) 1021–40.

42   S. Nadarajah and L. Vimalarajah, *The Politics of Transformation: The LTTE and the 2002–2006 Peace Process in Sri Lanka*, Transitions No. 4 (Berlin: Berghof Research Center for Constructive Conflict Management 2008).

43   Hocking (note 3) p. 359.

so. This is a crucial constraint in the Tamil liberation project: it is not simply a question of not being able to advocate armed struggle ('terrorism') against state repression, but also of not being able to promote the LTTE as the Tamils' political leadership *even when negotiating a political solution.* The point here is not whether the 'sole representative' claim is reasonable or not on its own terms, but whether it is legally possible to endorse it. This separation of 'Tamil grievances' from the LTTE is a key axiom of the global liberal order's approach to Sri Lanka's conflict. Despite the previous years of mass rallies, petitions and other articulations by European Tamils endorsing the LTTE as the Tamils' political representatives, the EU insisted, whilst banning the LTTE in 2006: '[this] decision is directed at the LTTE, and not at the Tamil people'.[44] Getting the *Tamils* to reject the LTTE as their political leadership and distance their political project from its armed struggle has long been an international objective, illustrated by the Australian government's declaration in November 1996 that it would only meet with Tamil groups 'provided they *condemn* in writing the terrorist activities of the LTTE'.[45] The proscriptions of the LTTE simply turn this preference into a legally enforceable requirement.

The (West-led) international community has thus, whilst demanding that Tamils reject the LTTE, assumed for itself the role of championing Tamils' rights vis-à-vis the Sri Lankan state (which it is also supporting against the LTTE). This self-nomination is also integral to the global liberal project, exemplified today by the logics of 'humanitarian intervention', 'responsibility to protect' and 'human security'.[46] The appropriate response for persecuted peoples is therefore not to take up arms against their state oppressors, but to call on the cavalry of the global liberal order for rescue. Thus, it is *global liberalism's* characterization of what constitutes the failings of the Sri Lanka state, rather than the Tamils' own, that comes to define the political terrain in which Tamil political activity can take place. Global liberalism defines the problem in Sri Lanka as 'a crisis of the state', of failures to meet international standards of governance, of institutional weakness, and so on.[47] Tamils also point to these, but argue, crucially, that these stem from the Sinhala-dominated state's institutional (constitutional, procedural and ideological) and thus insurmountable racism. Moreover, the Tamil *solution* is to exercise their right to self-determination; i.e. not to pursue *reform* of the majoritarian state, but to form a state of their own.[48] The foundation for the latter is a narrative that posits Tamils and Sinhalese as distinct nations, entitled to self-rule in their respective

---

44 Council of the European Union, 'Declaration by the Presidency on Behalf of the European Union Concerning Listing of the LTTE as a Terrorist Organisation' (Brussels 31 May 2006).

45 Comments by Australian Foreign Minister Alexander Downer to Parliament (5 November 1996), cited in an Australian Foreign Ministry letter dated 10 December 1996 to the Australian Federation of Tamil Associations. Emphasis added.

46 See for example Duffield (note 16); Richmond (note 13).

47 Goodhand (note 12) p. 30.

48 See Bose, Krishna (note 12).

homelands in the Northeast and South of the island. Thus, even when the 'solution' is framed in terms of 'autonomy' (rather than independence), the Tamil vision is very different to that of liberal peace. The former is based on recognition of the two nations and homelands and thus on sharing of power between them. However, the latter envisages a transformation of the present Sri Lankan state into a better governance structure, one explicitly privileging liberal and neoliberal values, rather than 'particularist' ones. Tamils may have 'legitimate grievances', but these must be addressed within a single, democratic, multiethnic space with no room for ethnic homelands. The individual (Sri Lankan citizen), rather than the (Tamil or Sinhala) nation, must thus be the unit and object of governance. With individual ('human') rights, rather than collective rights, thus coming to the fore, resolution of Sri Lanka's crisis is seen to turn primarily on ensuring of rule of law and equitable development (i.e. on ensuring economic opportunities for all individuals) rather than on recognizing the Tamils, as they demand, as a nation – that is, with collective and territorially grounded political rights – that requires protection from state repression. The need for 'autonomy' (say federalism) or 'devolution' is accepted, but only because it furthers the economic and political decentralization of the state required by liberal peace; there is, however, no room for recognition of 'ethnic homelands'.[49]

As the proscriptions of the LTTE have gradually extended across Western states, Tamil political activity in these countries has gradually come to focus less on endorsement of the LTTE as the Tamil *leadership* and more on abstract demands for 'peace talks' or 'a negotiated solution' involving the LTTE – a stance, on the face of it, not dissimilar to that of the international community. Similarly, Tamil advocacy increasingly turns less on the concepts of homeland, nationhood and self-determination, although these continue to inform Tamil politics, and more on specific *failings* – in liberal governmental terms, that is – of the Sri Lankan state's governance, including human rights abuses, inequitable allocation of state resources, crushing of media freedom, etc. For example, the killings in recent years by suspected security forces personnel or Army-backed paramilitaries of Tamil parliamentarians, journalists and political activists are constituted not as part of the state's efforts to crush the Tamil liberation project, but of the government's silencing of 'critics'. The Sri Lankan state's privileging of the Sinhala-dominated South and its exclusion of the Tamil-dominated Northeast in the allocation of international post-tsunami assistance[50] to Sri Lanka becomes 'inequitable'

---

49   For example, whilst many Tamil advocates of federalism envisage two federal states within Sri Lanka, recognising the Northeast as the Tamil homeland, some Western officials see up to ten states. Interviews 2006.

50   *TamilNet*, 'Tamil MP Slams Colombo for Ignoring Northeast Plight' (27 December 2004) at <http://www.tamilnet.com/art.html?catid=13&artid=13743>; *TamilNet*, 'Muslims demonstrate against Tsunami aid discrimination' (4 February 2005) at <http://www.tamilnet.com/art.html?catid=13&artid=14166>.

distribution, of inefficiency with ethnic overtones, rather than the wilful neglect of the non-Sinhala regions.[51]

Moreover, Tamil advocacy has increasingly come to demand the international community address these failings of liberal governance *rather* than to recognize state racism and the Tamil claim to self-determination. These elements are exemplified by the conduct of the Canadian Tamil Congress (CTC), a prominent diaspora lobby group known earlier for its advocacy for international recognition of the Tamil struggle for self-determination and of the LTTE.[52] Three months after Canada's 2006 ban on the LTTE, the CTC petitioned the government to 'appoint a *neutral* human-rights observer ... to document abuses *on all sides* [and] *throughout Sri Lanka*'.[53] The CTC also called on the Canadian government to 'appoint an *impartial* body to track aid flow into Sri Lanka *to ensure it is equitably distributed*'.[54] The CTC's efforts to avoid its earlier 'Tamil liberation' position hinges on its calls now for 'neutral' and 'impartial' intervention by Canada, rather than *in support* of the Tamils' struggle against oppression. Human rights abuses, rather than persecution of the Tamil nation, are posited as the problem for the Canadian government to address. The concern with abuses 'on all sides' and 'throughout Sri Lanka' fits with commonly stated international characterizations of the Sri Lankan crisis (that is, not of state persecution and consequent liberation struggle, but of a lack of rule of law and good governance) and of the international community (as concerned primarily with the welfare of Sri Lanka's residents, rather than self-interested support for the state).

The point here is the CTC is still seeking to promote the Tamil cause, but feels it cannot *only* argue, as it has done in the past, that in the context of the Sri Lankan state repression, Canada must support Tamil independence. To be seen as 'reasonable' or 'moderate' and not be dismissed (or worse, investigated) as 'LTTE-supporters', the CTC situates itself some distance from its earlier position of categorically demanding Tamil Eelam alongside the LTTE (which it now refers to as the present 'governing authority' – rather than 'representatives' – of the Tamils of Sri Lanka).[55] The CTC thus comes to echo the international discourse that posits both the LTTE and the Sri Lanka state as problematic. The CTC also seeks an agreeably reasonable 'impartial monitoring' of aid flow as opposed to an earlier stance by Tamil expatriates that the international community must support

---

51   For example, the report by the of the UN Special Envoy for Tsunami Reconstruction notes: 'What is particularly striking is the overprovision of house reconstruction in the south, which clearly indicates an inequitable allocation of resources from a national perspective' (November 2006) at <http://www.tsunami-evaluation.org/NR/rdonlyres/06B7033C-446F-407F-BF58-7D4A71425BFF/0/ApproachestoEquity.pdf> p. 21.

52   <http://www.canadiantamilcongress.ca/>.

53   Canadian Tamil Congress, 'Canadian Tamils Urge Three Actions for Peace' (Toronto, 25 July 2006). Emphasis added.

54   Ibid. Emphasis added.

55   Ibid. Emphasis added.

a joint aid-sharing mechanism between the LTTE and the Sri Lankan government. The inadvertent shift here accepts that the LTTE, like the Sri Lankan government, cannot be trusted with international aid – whereas a joint mechanism equates the legitimacy of both sides, impartial monitoring equates their illegitimacy. Furthermore the CTC's appeal now posits the Canadian government and the international community as the custodians of Tamil interests, rather than the LTTE or the Tamils themselves.

Conclusion

The central claim of this chapter has been that, by coercively shutting down specific political spaces and by providing alternative 'safe' spaces, the international anti-terrorism regime seeks to shape the Tamil diaspora's political activity in their hostlands towards realizing a liberal governmental vision for Sri Lanka. The point here is not that Tamil self-imagination as a persecuted people or an oppressed nation will be abandoned simply because the global liberal order wills it so. Rather, it is to argue that Tamils' political agitations increasingly come to take place in ways that reinforce the liberal order's problem–definition of Sri Lanka (i.e. a failure of governance requiring state reform), rather than reflect the Tamils' own sense of an oppressed nation seeking to exercise their right to self-determination.

The global liberal order has reconstituted the core issues confronting Sri Lanka's Tamils – such as institutionalized discrimination, embedding of Sinhala majoritarianism within the state bureaucracy and military, absence of physical security and rule of law, lack of media freedom, etc. – from being justifying elements of a demand for independence into targets of reform of the Sri Lankan state. These behavioural shifts are pursued on the basis of an asserted and implied conflation between the articulation of certain political positions (such as the demand for Tamil independence) and support for 'terrorist' violence (the LTTE's armed struggle). By setting out what are 'moderate' and 'extreme' views and simultaneously wielding the punitive apparatuses enabled by terrorism proscription, the global liberal order seeks to set the terms of what does and does not constitute acceptable advocacy.

This is not to say this sought after transformation of the Tamil liberation project has been achieved. Indeed, despite the increasing deployment of coercive anti-terrorism measures within Western states, expatriate Tamils continue to pursue their cause of national liberation from Sri Lankan state oppression in a variety of ways. However, amid the global liberal order's robust backing for the Sri Lankan state in confronting the LTTE, they are compelled to do so with considerable care and guile to avoid the terrifying apparatuses of domestic security enabled by anti-terrorism legislation. Interestingly, in recent years, the Tamil liberation project has been increasingly reinforced by the Sri Lankan state's own deepening resistance to the global liberal order. The strident Sinhala nationalism and chauvinism that has engulfed the Sri Lankan state and polity since 2004, particularly as the internationally-backed military campaign against the LTTE has gained ground,[56] has made international assertions of the reformability of the Sri Lankan state along liberal lines increasingly untenable. Indeed, it is amid a manifest inability or unwillingness of the global liberal order to discipline and transform the Sri Lankan *state* that the call for an independent Tamil Eelam has emerged forcefully again at the centre of Tamil expatriate agitation. Notably, however, even this long-standing goal of independence is now pursued by mobilizing the referential terms of liberal governmentality alongside, if not ahead of, the foundations of nation, homeland and self-determination that formed the basis for it before. To say that terrorism proscription has thus far failed to transform the Tamil liberation project into one of liberal state-building in Sri Lanka is not, therefore, to deny the potency of the securitization of politics that the terrorism discourse constitutes.

---

56   D. Rampton with A. Welikala, 'Colliding Worlds: Sinhala Nationalism and Populist Resistance to the Liberal Peace', in J. Goodhand, J.B. Korf and J. Spencer (eds), *Caught in the Peace Trap? Conflict, Aid and Peacebuilding in Sri Lanka* (Routledge, forthcoming).

Chapter 7

Negotiating Security: Governmentality and Asylum/Immigration NGOs in the UK

Patricia Noxolo

Introduction

The climate of insecurity following the terrorist attacks on New York on 11 September 2001 and in the UK following the terrorist attacks on London transport on 7 July 2005 has been met by strategic responses from many different kinds of actors. This chapter considers how UK-based non-governmental organizations (NGOs) have negotiated the securitization of immigration and asylum by examining the networking strategies and scalar politics that they have employed.

In order to explore this question, the chapter examines the discursive repositioning of a range of NGOs that have either a main or subsidiary concern with migration and asylum.[1] It draws on research employing two main methods. The first is discursive analysis of NGO documents published since 2001, mainly taken from their websites. These documents take a variety of forms and are aimed at a variety of audiences, including newspaper editorials and press releases for publication in the mainstream press; written evidence and briefings for parliamentary committees and judicial bodies; and information sheets, reports and briefings for wider audiences. The second method is interviews with representatives of a range of UK-based NGOs, including the Immigration Law Practitioners Association (ILPA), Institute for Race Relations (IRR), Joint Council for the Welfare of Immigrants (JCWI), Justice, Migrationwatch UK and the Refugee Council.

---

1   The choice of NGOs reflects a broad definition of these organizations in relation to free movement – organizations that operate within the UK, but that are directly affiliated neither to the UK government nor to any particular for-profit enterprise. This definition obviously covers an enormous number of organizations (see J. Kendall and M. Knapp, *The Voluntary Sector in the UK* [Manchester: Manchester University Press 1996]) engaged in a wide range of activities. I am not seeking to erect a typology of NGO organizations, particularly as this is notoriously difficult to do. However, I have chosen these organizations as important players in relation to several key NGO functions surrounding migration and asylum: promoting human rights and civil liberties; supporting refugee and asylum groups; supporting minority groups; representing employers and immigration professionals; campaigning for employment rights; campaigning for increased immigration controls; and campaigning against racism and discrimination.

In recent years, most notably since the inauguration of the global war on terror, many writers have observed that global security concerns have changed both in shape and in intensity,[2] entering many different dimensions of political and economic life,[3] and in terms of a range of new or transformed actors now involved in security practices.[4] Didier Bigo has combined these insights with a Foucauldian governmentality analysis, arguing that, in Europe at least, immigration has been placed within 'a continuum of threats and general unease'.[5] Thus, asylum and immigration now form part of a constellation of problematic issues (including crime and terrorism) that are seen as inherently interlinked and which are commonly viewed through the prism of security. Such a situation has come about not as a result simply of 'speech acts' by particular governmental actors, but by the consonance of global structural trends with the legitimating moves of a range of political actors and the technological apparatus and expertise of transnational security personnel:

> Securitization, then, is generated through a confrontation between the strategies of political actors (or of actors having access to the political stage through the media), in the national political field, the security professionals at the

---

2   See R. Abrahamsen, 'A Breeding Ground for Terrorists? Africa and Britain's "War on Terrorism"', *Review of African Political Economy* 31/102 (2004) 677–84; M. Duffield, 'Reprising Durable Disorder: Network War and the Securitization of Aid', in B. Hettne and B. Oden (eds), *Global Governance in the 21st Century: Alternative Perspectives on World Order* (Stockholm: Almkvist & Wiksell International 2002); R. Higgott, *American Unilateralism, Foreign Economic Policy and the 'Securitization' of Globalization* (Singapore: Institute of Defence and Strategic Studies 2003).

3   S. Elbe, 'The Futility of Protest? Biopower and Biopolitics in the Securitization of HIV/AIDS' (paper presented at the ISA Convention, Montreal, Québec 2004); J. Eriksson, 'Cyberplagues, IT, and Security: Threat Politics in the Information Age', *Journal of Contingencies and Crisis Management* 9/4 (2001) 211–22.

4   S. Bislev, D. Salskov-Iversen and H. Krause Hansen, 'Governance and Globalization: Security Privatization on the US-Mexican Border: A New Role for Non-state Actors in Security Provision?', *IKL Department of Intercultural Communication and Management Working Paper* 42 (2001); A. Leander, 'Privatizing the Politics of Protection: Military Companies and the Definition of Security Concerns', in J. Huysmans, A. Dobson and R. Prokhovnik (eds), *The Politics of Protection: Sites of Insecurity and Political Agency* (London: Routledge 2006) 19–33; O. Waever, 'The EU as a Security Actor: Reflections from a Pessimistic Constructivist on Post-sovereign Security Orders', in M. Kelstrup and M. Williams (eds), *International Relations Theory and the Politics of European Integration: Power, Security and Community* (London: Routledge 2000) 250–94.

5   D. Bigo, 'Security and Immigration: Toward a Critique of the Governmentality of Unease', *Alternatives* 27 (2002) 63–92, p. 63; M. Foucault, 'Governmentality', in G. Burchell, C. Gordon and P. Miller, *The Foucault Effect: Studies in Governmentality* (Hemel Hempstead: Harvester Wheatsheaf 1991) 87–104.

transnational level (public and private bureaucracies managing the fear), and the global social transformations affecting the possibilities of reshaping political boundaries.[6]

The security prism thus gives rise to a form of globalized governmentality, operating through but also across national, regional and global scales.

Many writers have specifically commented on the ways in which NGOs have become linked into increasingly complex relationships with governments and with the art of government. Rose and Miller have shown that alliances between government and NGOs are indispensable to the operation of government at a distance, and more recently Morrison has noted the development of compacts through which the UK government and the voluntary sector are working out joint rationalities for governing the national population.[7] Dean has emphasized the technologies of agency and performance by which NGOs become part of a system of self-managing units 'in which the regulation of services and the management of budgets is undertaken by the polymorphism of the audit and various kinds of accounting'.[8] Many see the work that NGOs do with client groups as the operation of techniques of the self, 'normalizing' or 'responsibilizing' (potential) citizens, and transforming people into governable subjects.[9]

This view of NGOs as linked in with the arts of government is one of a range of critical perspectives on NGOs in recent years, which recognize their changing roles and conditions under neoliberal forms of globalization;[10] however viewing NGOs in terms of governmentality does not necessarily preclude attention to the

---

6   Bigo (note 5) p. 75.

7   J. Morison, 'The Government-voluntary Sector Compacts: Governance, governmentality, and Civil Society', *Journal of Law and Society* 27/1 (2000) 98–132; N. Rose and P. Miller, 'Political Power Beyond the State: Problematics of Government', *The British Journal of Sociology* 43/2 (1992) 172–205.

8   M. Dean, *Governmentality: Power and Rule in Modern Society* (London: Sage 1999) p. 170.

9   R. Bryant, 'Non-governmental Organizations and Governmentality: 'Consuming' Biodiversity and Indigenous People in the Philippines', *Political Studies* 50 (2002) 268–92; S. Ilcan and T. Basok, 'Community Government: Voluntary Agencies, Social Justice, and the Responsibilization of Citizens', *Citizenship Studies* 8/2 (2004) 129–44; W. Larner and W. Walters, 'Privatization, Governance and Identity: The United Kingdom and New Zealand Compared', *Policy and Politics* 28/3 (2000) 361–77.

10   W. DeMars, *NGOs and Transnational Networks: Wild Cards in World Politics* (London: Pluto 2005); D. Lewis, 'Individuals, Organizations and Public Action: Trajectories of the 'Non-governmental' in Development Studies', in U. Kothari (ed.), *A Radical History of Development Studies: Individuals, Institutions and Ideologies* (London: Zed 2005) 200–222; J. Murphy, 'The World Bank, INGOs and Civil Society: Converging Agendas? The Case of Universal Basic Education in Niger', *Voluntas: International Journal of Voluntary and Nonprofit Organizations* 16/4 (2005); T. Wallace, 'Trends in UK NGOs: A Research Note', *Development in Practice* 13/5 (2003).

ways in which NGOs can be resistant to aspects of state control.[11] As Barnett has pointed out in the context of cultural geography, Foucauldian governmentality does not presuppose a unitary and all-encompassing state domination (within which NGOs might be seen as simply incorporated): rather governmentality is concerned with the interactions between different discursive frameworks emanating from subjects who are governed through their freedom, that is with 'the heterogeneity of the discourses and practices that shape any particular strategy or apparatus of government.'[12]

Networking is increasingly important to NGOs as a way of using scarce resources more effectively, by pooling influence[13] and knowledge.[14] Networking at and between a range of scales, NGOs can now be seen as multi-scalar networked actors, networking 'vertically' (for example local community organizations networking with national, regional and global umbrella organizations), and 'horizontally' (with other similar organizations through joint forums and meetings).[15] NGOs not only network with each other, but also with government representatives and within intergovernmental systems of governance, plus a range of other professionals (for example welfare, legal and security professionals).[16] Through networking NGOs 'jump scales', working with policymakers far beyond the local scale, in order to make policy interventions, share information and provide services.[17]

Beyond this, through this combination of increasing links with the arts of government and creative multi-scalar networking, NGOs should also be seen as deeply involved in *constructing* a range of scales. The concept of scale as socially constructed is one that has become more or less commonplace in geography in recent years, particularly in the context of the changing spatiality of state-

---

11   A. Bebbington, S. Hickey and D. Mitlin, 'Introduction: Can NGOs Make a Difference? The Challenge of Development Alternatives', in A. Bebbington, S. Hickey and D. Mitlin (eds), *Can NGOs Make a Difference? The Challenge of Development Alternatives* (London: Zed 2008) 3–38, p. 8.

12   C. Barnett, 'Culture, Government and Spatiality: Re-assessing the "Foucault Effect" in Cultural-policy Studies', *International Journal of Cultural Studies* 2/3 (1999) 369–97, p. 384; see also M. Foucault, *Sécurité, Territoire, Population: Cours au Collège de France 1977–1978* (Paris: Gallimard Seuil 2004).

13   H. Yanacopulos, 'The Strategies that Bind: NGO Coalitions and their Influence', *Global Networks* 5/1 (2005) 93–110.

14   J. Bach and D. Stark, 'Link, Search, Interact: The Co-evolution of NGOs and Interactive Technology', *Theory, Culture and Society* 21/3 (2004) 101–17.

15   DeMars (note 11).

16   P. Willetts, 'What is a Non-Governmental Organization?', in *UNESCO Encyclopaedia of Life Support Systems: Section 1 Institutional and Infrastructure Resource Issues* (2002) at <http://www.staff.city.ac.uk/p.willetts/CS-NTWKS/NGO-ART.HTM>.

17   See for example T. Perreault, 'Changing Places: Transnational Networks, Ethnic Politics, and Community Development in the Ecuadorian Amazon', *Political Geography* 22 (2003) 61–88.

centred governmentality.[18] However, placing scalar construction within a broadly Foucauldian governmentality framework focuses attention not only on power as operated by the state but on the diffuse operation of power in and through a range of political subjects and their interactions.[19] Local, regional, national and global scales are thus understood as produced *relationally*, not only in relation to each other, but through the operations of power within the myriad interactions and negotiations between a range of groups, including civil society, politicians and citizens. At the same time, a Foucauldian framework highlights the role of discourse and the politics of knowledge in the operations of power, focusing not only on material practice, but also on language and other forms of representation. Attention to the 'politics of geographical knowledge and the power of geographical representation' has also become an accepted part of critical geopolitics, and these post-structural forms of analysis can usefully be brought to bear on NGOs, particularly in the aspects of their work that have to do with campaigning and fundraising.[20] As NGOs engage increasingly with public relations/advertising and with research/information-sharing, the politics of representation has become more and more central to their work.[21]

---

18   N. Brenner, 'The Limits to Scale? Methodological Reflections on Scalar Structuration', *Progress in Human Geography* 15 (2001) 525–48; S. Marston, J.P. Jones and K. Woodward, 'Human Geography without Scale', *Transactions of the Institute of British Geographers* 30 (2005) 416–32; Neil Smith, 'Homeless/global: Scaling Places', in J. Bird, B. Curtis, T. Putnam, G. Robertson and L. Tickner (eds), *Mapping the Futures: Local Cultures, Global Change* (London: Routledge 1993) 87–119; N. Smith, 'Spaces of Vulnerability: The Space of Flows and the Politics of Scale', *Critique of Anthropology* 16 (1996) 63–77. Regarding governmentality, see T. Bunnell and N. Coe, 'Re-fragmenting the 'Political': Globalization, Governmentality and Malaysia's Multimedia Super Corridor', *Political Geography* 24 (2005) 831–49; B. Jessop, 'The Changing Governance of Welfare: Recent Trends in its Primary Functions, Scale and Modes of Coordination', *Social Policy and Administration* 33 (1999) 348–359.

19   M. Foucault, *Sécurité, Territoire, Population: Cours au Collège de France 1977–1978* (Paris: Gallimard Seuil 2004); M. Foucault, 'Two lectures', in C. Gordon (ed.), *Michel Foucault: Power/Knowledge: Selected Interviews and Other Writings 1972–1977* (New York: Harvester Wheatsheaf 1980).

20   K. Dodds, 'Political Geography III: Critical Geopolitics After Ten Years', *Progress in Human Geography* 25/3 (2001) 469–84, p. 470. See also G. Ó Tuathail, *Critical Geopolitics: The Politics of Writing Global Space* (Minneapolis: University of Minnesota Press 1996); D. Slater, *Geopolitics and the Post-Colonial: Rethinking North-South Relations* (Oxford: Blackwell 2004).

21   On communications see D. Deacon, 'The Voluntary Sector in a Changing Communication Environment: A Case Study of Non-official News Sources', *European Journal of Communication* 11/2 (1996) 173–99; H. Lidchi, 'Finding the Right Image: British Development NGOs and the Regulation of Imagery', in T. Skelton and T. Allen (eds), *Culture and Global Change* (London: Routledge 1999) 87–101. On networking see J. Bach and D. Stark, 'Link, Search, Interact: The Co-evolution of NGOs and Interactive Technology', *Theory, Culture and Society* 21/3 (2004) 101–117; H. Holmen, *NGOs,*

In this context, as Marston et al. point out, the question of a *politics* of scale takes at least two different forms: it can be understood in terms of the ways in which political agents' spatial practices extend or transform the conventions of political relations and practices at or between scales (e.g. practices of 'scale jumping' or 'scale bending');[22] however, a politics of scale can also be understood as 'the need to expose and denaturalize scale's discursive power', i.e. the need to analyse the ways in which the discourse of scale is represented and deployed as a technology by a range of actors within an ongoing politics of control and negotiation.[23] In the context of the securitization of immigration therefore, the scalar politics of NGOs can be understood in two ways: in relation to the production of scale through the networking practices of NGOs; and in relation to the production of scale through the spatial representations constructed by NGOs in their publications.

Before exploring each of these aspects in turn, it must be emphasized that NGOs primarily concerned with immigration and asylum deal with a range of issues, most of which do not concern security: in fact some were clear that to deal specifically with these issues might be seen as a diversion from normal work. These organizations often have a very small number of full time staff (typically between four and ten people), and although some have had to acquire additional skills in relation to security, interviews revealed that they have not on the whole acquired staff dedicated to this issue. The same is true for NGOs for whom immigration has been relatively marginal. They engage with security issues generally on their primary terrain, for example civil liberties or representing the range of employer or professional interests – on the whole they have not changed the structure of their organization specifically to take on immigration staff. It is important to highlight this because the connection between security and migration is neither necessary nor obvious: it is highly contingent, the effect of particular discursive framings in a particular historical moment.

Negotiating the Securitization of Immigration: Networking as Scalar Politics

All the organizations interviewed engage with networking of various kinds in the general course of their work, both within the UK and internationally. Some organizations were instituted as part of an international federation of organizations (for example, Justice was formed in relation to the International Commission of

---

Networking and Problems of Representation (Linkopings University and ICER 2002) at <http://www.icer.it/docs/wp2002/holmen33-02.pdf>. On the politics of representation see also M. Duffield, *Global Governance and the New Wars: The Merging of Development and Security* (London: Zed Books 2001); S. Robins, 'NGOs, "Bushmen" and Double Vision: The ≠ Khomani San Land Claim and the Cultural Politics of "Community" and "Development" in the Kalahari', *Journal of Southern African Studies* 27/4 (2001) 833–53.

22   Marston et al. (note 19).
23   Ibid., p. 420.

Jurists, and has counterparts in many Commonwealth countries), whilst most have formal or informal European and other international links. For example IRR have an EU officer, whose role frequently includes speaking in European forums on a range of issues; she was clear that there has been a great deal of interest in IRR's publications on deportation and on securitization, and this has enhanced this kind of informal networking opportunity.[24]

The nexus between immigration/asylum and security has engendered a great deal of work for NGOs in terms of monitoring and assessing the impact for asylum seekers and immigrants of changes in UK security legislation (such as the Anti-Terrorism, Crime and Security Act 2001), as well as numerous changes in immigration legislation, and the impact within the UK of EU directives.[25] Most of the representatives expressed the view that networking has been a way to pool resources in order to do this effectively, and liaison with NGOs specifically monitoring EU debates around asylum/immigration and security has been an important source of information.[26] This said, however, the representative from ILPA stated that anti-terrorism measures are being worked out separately in each individual country. The roles of NGOs in monitoring and responding to changes in legislation, including frequent participation in the parliamentary committee debates surrounding these changes, has therefore meant that networking at the national scale has been extremely important.

One strategy for NGOs concerned directly with immigration/asylum has been to use networking at the national scale as a way to address the British government's security initiatives without having to become too clearly identified with security per se. Interviews with Refugee Council and JCWI revealed a cautious attitude to the politics of security, with a wariness that talking about security and immigration together may make the link between them seem more obvious rather than less so. As the representative from Refugee Council succinctly put it: '*by responding to this link there is a risk that you can perpetuate it*', that is, the link between asylum seekers and terrorism may come to seem self-evident, rather than contingent. The representative from JCWI stated that they network with organizations who are more high profile in relation to challenging security measures, groups such as NO2ID (who are campaigning against the introduction of identity cards), as well as Justice and Liberty who have made high profile contributions to debates over civil liberties. Through this networking they are able to give a relatively high national profile to the ways in which immigration policies are affected by anti-

---

24   L. Fekete, 'All in the Name of Security', in P. Scraton (ed.), *Beyond September 11: An Anthology of Dissent* (London: Pluto Press 2002) 102–107; L. Fekete, *The Deportation Machine* (London: Institute of Race Relations 2005).

25   See D. Flynn, 'New Borders, New Management: The Dilemmas of Modern Immigration Policies', *Ethnic and Racial Studies* 28/3 (2005) 463–90 for a summary of relevant changes.

26   See for example Statewatch <http://www.statewatch.org/> and European Council on Refugees and Exiles <http://www.ecre.org>.

terrorism legislation (for example the increased risks of indefinite detention or deportation), whilst ensuring that these issues are addressed as human rights issues in the context of security concerns. This manoeuvre allows these organizations to challenge the heightened context in which security issues become linked with immigration, without itself appearing to endorse the idea that immigrants and asylum seekers are uniquely or generically linked with terrorism.

By the same token, networking around public communication, particularly in the form of the joint statement, has become increasingly important as a way for organizations to challenge the links between security and migration in a less direct way, more distanced from other institutional aspects of their work. For example, in the 'Joint Response to the Home Office consultation on exclusion or deportation from the UK on non-conducive grounds', August 2005, to which a variety of British NGOs (The Refugee Council, Refugee Action, Immigration Advisory Service, Oxfam GB, Scottish Refugee Council, Welsh Refugee Council, Amnesty International UK, The Medical Foundation) were signatories, the words 'community', 'communities' and 'society' were able to shift and change, without any of the imperatives to pin down their meaning that might come from the needs of individual NGOs to legitimate themselves.[27] This shifting has the effect of blurring the boundaries between the concerns of citizens and non-citizens of the UK, allowing for potential differences in relation to criteria for security and insecurity whilst never settling on the kind of formulation that would offer complete closure as to the identity and membership of particular groups ('the British public' or 'UK society'). Such tactics are evident in the following passage from this document:

> Careful consideration is needed if the security of communities in the UK is to be achieved without generating feelings of insecurity and alienation among some sections of society.

> 1.3 In recent weeks, we have become aware of an enhanced sense of fear and insecurity among some refugee communities and others subject to immigration controls. For some this fear is linked to the general threat of terrorism faced by the UK. For others it stems from the recent rise in racial and religious attacks. And for some it is linked to a fear that they themselves may be unduly and unfairly affected by counter-terrorism measures and returned to unsafe situations in their country of origin.

> 1.4 There is a real need for widespread consultation and public information to ensure that communities' fears are acknowledged and addressed. At a time of great tension it is particularly important that no group feels scapegoated because of the actions of particular individuals. Ongoing engagement with all

---

27   S. Lister, 'NGO Legitimacy: Technical Issue or Social Construct', *Critique of Anthropology* 23/2 (2003) 175–92.

communities in the UK is also essential in gathering the right intelligence to combat terrorism.

As one Refugee Council representative put it in an interview with me:

> Security needs to be seen as a continuum, from fear of exclusion and lack of services to terrorism. Refugee communities would therefore be understood as more at risk because they can't access services.

The scalar politics of NGOs around the securitization of immigration and asylum can be seen as strategic use of networking at the national scale in order to directly address national government initiatives without reinforcing the contingent connections between security and immigration/asylum. Regional and global networks remain important sources of information and support, however. Even this careful strategy needs to be used with caution though, as networking may itself have the effect of institutionalizing links between security and immigration/asylum. The representative from Justice expressed some concern that, in the context of a heightened politics around security, an immigration/security nexus can in and of itself be a point of convergence between disparate NGOs. As Bigo pointed out in relation to security personnel, the presumed relationship between immigration and security can become a *lingua franca* for a number of different organizations who want to work together. There is a risk that the institutional and political pressure to network may lead the connection between immigration and security to seem gradually less contingent and increasingly 'obvious' to these organizations. This may make it more difficult to challenge in the long run.

Representation and Counter-Representation as Scalar Politics

Scalar politics is evident in the strategies of NGOs around asylum/immigration and security in a variety of ways. However, not every NGO is concerned primarily with the well being or interests of migrants or asylum seekers. In January 2003, during some of the worst crises around asylum/immigration and security, the anti-immigration NGO Migrationwatch UK engaged in a representational scalar politics that was focused on the erection of boundaries around the nation state and the abjection of terror to the spaces beyond that boundary. In this representational framework, the nation therefore becomes the physical container of safety and protection, whilst the international scale, with its international refugee convention and its global movements of migrants, was figured as a direct threat to that sense of safety and protection.

Pro-immigration/asylum NGOs had difficulty responding effectively to this because every intervention they made ran the risk of perpetuating the connection between asylum seekers and terrorism which was being generated in the public discourse. This is evidence of the ways in which securitization as a practice of governmentality structures the range of interventions possible within the field, meaning that the range of discursive strategies open to NGOs become limited: 'Even when NGOs intervene, they can do so only by turning professional, by producing this kind of knowledge'.[28] The response of more liberal NGOs has been to represent the global scale as the only space at which security concerns can effectively be addressed, thus directly countering the logic of defensive localism in the anti-immigration stance. This section examines this episode in more detail.

Migrationwatch UK, established in December 2001, describes itself as an 'independent think tank which has no links to any political party or organization'.[29] It has also been described as 'a pressure group with a distinctly unpleasant agenda'.[30] They are a membership organization whose objective is to reduce flows of immigration into the UK, which they picture as too high:

> We are not opposed to immigration that is moderate and managed. At present it is neither. We wish to ensure that the arguments adduced in favour of the current large-scale immigration are thoroughly examined as we believe them to be unsound. We also believe that such massive immigration is contrary to the interests of all sections of our community, adding to the problems of both overcrowding and integration.[31]

These arguments are certainly not new. There is a long history of alarms and arguments over numbers as a strategy for limiting migration, dating back at least to the 1950s Conservative governments of Macmillan and Eden.[32] Other arguments relating to the threats posed by migration are also not new, and are also not in general clearly related to terror. The threats raised by the group take multiple and changing forms: that the numbers of legal immigrants are too high; that failed asylum seekers are not removed; that the numbers of people entering the country illegally are unknown and that departures are not recorded; that new migrants carry diseases into the UK (particularly HIV/AIDS); that illegal immigrants and failed asylum seekers obtain work and health benefits without making a contribution to tax; and that the government has lost control of population and immigration figures. Each of these assertions is constantly stated and re-stated in multiple forums (newspapers; parliamentary groups; and the group's website). Each is 'supported' by demographic analysis from the group's own analysts.

Because Migrationwatch UK was established just after the attacks of 11 September 2001, it is possible to see the heightened global tension surrounding

---

28   Bigo (note 6) p. 83.

29   <http://www.migrationwatch.com>.

30   *The Independent* (6 August 2002).

31   <http://www.migrationwatch.com>.

32   B. Carter, C. Harris and S. Joshi, 'The 1951–55 Conservative Government and the Racialization of Black Immigration', in W. James and C. Harris (eds), *Inside Babylon: The Caribbean Diaspora in Britain* (London: Verso 1993) 55–73.

security as the midwife to its birth as a mainstream political voice. The founder member of the organization, Sir Andrew Green, a former British diplomat in Saudi Arabia, began to look at issues of asylum and migration after the failure of the British government to remove Mohammed al-Masari, a dissident Saudi physicist whom he described as an 'Islamic extremist'.[33] Though the group clearly has political connections and resources, it was not until late 2001 that the group began to gain significant public attention with its projections of migration figures. Dr. David Coleman, the group's resident demographer, 'said he was taken aback by the scale of the response to the figures that he had put together for the group. "I have been writing about this for some time but no one has really latched on to it", he said'.[34]

The group continued to issue regular press releases in the months that followed on a range of anti-immigration issues. However, it was not until January 2003 that the group began to talk explicitly about security. This was three months after the October 2002 Bali bombings, and in the run-up to the war on Iraq, which began in March 2003. Explicit links between asylum seekers and terror were also being made in the press in the wake of the ricin plot, in which PC Stephen Oake was killed while carrying out a raid on the home of a suspect terrorist and in which some suspects were said to be asylum seekers.[35]

At this time Migrationwatch UK placed several press releases in which they explicitly related asylum to security concerns, calling for detention of asylum seekers who could not prove their identities, and calling for entitlement cards as a way of cutting illegal immigration, both of which are represented as security measures. In a press release of 15 January 2003, 'Entitlement cards now urgent priority', Migrationwatch UK, after talking about what they considered the lack of proper record keeping about who comes and goes, added: 'There are also significant consequences for security ... No single measure will defeat the terrorists but we

---

33   *Independent* (note 30).

34   Ibid.

35   See for example *The Sun*: 'Asylum in Britain is Now a Trojan Horse for Terrorism', 23 January 2003 and M. Phillips in the *Daily Mail*: 'the asylum shambles is the sea in which terror most easily swims', quoted in 'Asylum Issues Lead News' (20 January 2003) at <http://news.bbc.co.uk/1/low/uk/2675427.stm>. For more general analysis of the treatment of refugees and asylum seekers in the UK media, C. Coole, 'A Warm Welcome? Scottish and UK Media Reporting of an Asylum Seeker Murder', *Media, Culture and Society* 24 (2002) 839–52; ICAR, *Media Image, Community Impact* (London: Information Centre about Asylum and Refugees in the UK 2004); M. Malloch and E. Stanley, 'The Detention of Asylum Seekers in the UK: Representing Risk, Managing the Dangerous', *Punishment and Society* 7/1 (2005) 53–71; K. Wells and S. Watson, 'A Politics of Resentment: Shopkeepers in a London Neighbourhood', *Ethnic and Racial Studies* 28/2 (2005) 261–77; A. White, 'Teaching Students to Read the News via Representations of Asylum Seekers in British Newspapers', *Journal of Geography in Higher Education* 28/2 (2004) 285–99. On the 'ricin plot' see BBC News, *Four Cleared of Poison Conspiracy* (13 April 2005) at <http://news.bbc.co.uk/1/hi/uk/4441993.stm>.

must tilt the balance against them'. Again on 16 January 2003, 'Security now paramount consideration', the organization put out a press release that focused on security:

> The increased terrorist threat to the UK leaves little alternative but to detain those asylum seekers who destroy their documents until their identities have been established and security checks made, says independent think-tank MigrationwatchUK … "While the vast majority of asylum seekers are law-abiding people recent events have graphically demonstrated that they include a small, but very dangerous minority", said MigrationwatchUK Chairman, Sir Andrew Green.

At a time when national, European and global attention was focused on terror, Migrationwatch UK were able to represent the national scale as a space where security could be maintained through the monitoring and detention of asylum seekers. In other words, terror was depicted as something that permeates into the local from the global scale, through Britain's commitment to the international refugee convention.

Unlike Migrationwatch UK, the majority of NGOs concerned with migration and asylum are broadly supportive of more liberal immigration and asylum measures, and are critical of the implication that more draconian measures in relation to asylum seekers will improve security. However, an extended discursive analysis of Refugee Council's immediate response to the January 2003 moral panic around asylum and terror, which took the form of a briefing, entitled 'Asylum and terrorism – the facts', published on 21 January 2003, illustrates Bigo's point about the difficulty of responding to the logic of abjection in a securitized context. The briefing began with a rational countering of rumour and hype with facts and figures:

> The figures speak for themselves. 88,300 people applied for asylum in the UK in 2001. Of the thirteen people currently being held following the police activity in London and Manchester last week, it is now being alleged that two entered the country via the asylum system … A number of other people who were initially taken into custody, some of whom were purported to be asylum seekers, have subsequently been released without charge. It is also currently alleged that one of the seven people detained following the raid on the Finsbury Park mosque on Monday had applied for asylum. This makes a total of three asylum seekers reportedly held by police in connection with recent events. The association of asylum seekers en-masse with terrorism is wildly misleading and irresponsible.

The briefing then went on to use a particular strategy to counteract the linking of terror and asylum: it reinstates asylum as a symbolic drama, with three protagonists – the asylum seeker/refugee (who, the document reminds us is much more likely to be a victim than a perpetrator of violence), the UK, and oppressive governments elsewhere from whom people need to seek asylum. The terrorist is inserted into this drama as a fourth protagonist who comes into the UK from the outside, but not within the asylum system. The terrorist is pictured as well-resourced, with criminal networks which, crucially, enable him/her to take advantage of open borders, in contrast with vulnerable asylum seekers:

> It would be impossible to make our borders utterly impervious to criminals and terrorists – dangerous people who have the money and networks to get in to the UK without recourse to the asylum system. But it would be morally inexcusable to abandon our commitment to provide protection to those who need our help.

The last two paragraphs of the document bring the drama together with reference to the 1951 Convention Relating to the Status of Refugees:

> The right to protection is enshrined in international law in the 1951 United Nations Convention Relating to the Status of Refugees. Under the terms of the Convention, there should be no safe haven for terrorists, nor are they protected from criminal prosecution if they transgress against the laws of their host country. On the contrary, the Convention is carefully framed to exclude people who have committed particularly serious crimes, whilst ensuring sanctuary to those with a genuine need.

> The statistics show that the majority of people who seek refuge in the UK have taken flight from regimes with well-documented and incontestable records of human rights abuses – from countries like Iraq, Zimbabwe, Somalia and Afghanistan. The UK has a strong tradition of giving people uprooted by war and injustice the chance to rebuild their lives – and over the years we have benefited enormously from the skills refugee communities have brought to our shores. We must not falter in our support for the world's most vulnerable now.

Here the convention provides an international legal framework from which terrorists are excluded, in order to deal with oppressive regimes in localities dotted around that international space, abusing human rights. This tension and danger at the local scale (in other locales) is resolved through a representation of the UK as a local space in which individuals from around the world can find safety and be given a chance. An interesting scalar politics is set up during this drama therefore: the national scale (in other nations) is a source of oppression, and the international scale is a source of conventions designed to mitigate that oppression. However, in contrast, the UK is a space at the local scale where international law is put into action and there is safety for the 'world's most vulnerable'. Within the asylum drama then, international and national scales interact, the international mitigating the problems of the national, the national implementing the laws of the international.

At the same time, however, the document struggles in a securitized context to avoid identifying local–global relations outside the asylum regime as the main source of danger. As part of its rational depiction of facts and figures, the document emphasizes the large numbers of cross-border movements by people who are not asylum seekers, though carefully avoiding directly saying that terrorists are more likely to be business people, tourists or students:

> We should also keep in mind that the number of people who enter the country and seek protection through the asylum system is minimal when considered in the context of the annual traffic across UK borders. During 2001, around 23 million individuals (including business people, tourists and students) came from abroad to stay for a limited period in the UK. 108,825 people came to stay on work-permits, and 29,100 people came to the country to rejoin their husband or wife. In total, over 88 million passed through UK borders during the course of the year.

However, by referring to the broader phenomenon of migration, in contrast to asylum, and the impossibility of strengthening borders in the context of security threats, the document does suggest that the terrorist threat is linked with the opening of borders that constitutes the global scale, and gets caught in a rationality of abjection of danger that does not effectively challenge Migrationwatch's defensive scalar politics.

Those NGOs concerned more with human rights as such, have tended to offer a more direct challenge to the defence of the local against the global, by representing international principles as a source of protection against national partiality, for example in the opposition between the upholding of universal principles of human rights – for example the right to a fair trial – against national security issues, which may appear to be furthered by deportation. In 2006, Amnesty UK criticized the UK government over the impact of its anti-terror measures on global standards of human rights, particularly around memoranda of understanding that were meant to protect people subject to deportation from abuse, by the authorities in the country to which they were being deported:

> As a Permanent Member of the UN Security Council, the current President of the EU, and member of the G8, Council of Europe and the OSCE, the global impact of the UK's actions and proposed measures cannot be overstated. In addition, the credibility of the UK as an advocate of human rights abroad is also rapidly eroding due to its own demonstrated lack of respect for human rights.[36]

The corollary of this is the representation of a commitment to universal human rights as the most effective way to increase both global and local security. This

---

36   Amnesty International UK, *Counter-Terrorism and Human Rights in the UK* (2006) at <http://www.amnesty.org.uk/actions_details.asp?ActionID=93>.

is done by insisting on the portrayal of the terrorist as an international criminal with access to globalized technology, which makes deportation ineffective or irrelevant as a security measure. For example, in August 2005 Liberty questioned why suspected terrorists must be deported to places with poor human rights records, arguing instead that the universal principle of crime and punishment should be invoked: 'Ultimately, terrorists should be charged, tried and convicted, not shuffled off around the world only to return, or peddle their hatred elsewhere.' From this position of looking at global security as linked with universal rights, Liberty challenge the link between anti-terrorism and immigration control as a political device that distracts from the problem at hand, since British citizens are as vulnerable to radicalization as people in any other part of the world:

> we are disappointed in the repeated suggestion that combating terrorism is a matter of immigration control. Peddling this line may play well in parts of the country but ducks the more serious and difficult questions about how terrorist organizations are able to successfully recruit amongst Britons and non-Britons alike.

Perhaps it is not surprising that there should be less emphasis on deportation following the attacks in London on 7 July 2005, where much was made of the fact that the four men involved were 'home-grown' terrorists, neither asylum seekers nor recent migrants. Since then the Refugee Council have made a number of other interventions around the securitization of asylum which cut across the logic of the UK as necessarily a safe haven, for example initiating research on the impact of anti-terror measures and the climate of fear surrounding refugees and asylum seekers, and in a reactive way, by making critical contributions to parliamentary debates as expert witnesses or as commentators.[37]

Conclusion

This chapter has assessed the ways in which UK-based NGOs have engaged in networked scalar politics in relation the securitization of immigration. Through the discourse analysis of a range of published materials and interviews, the chapter has shown that networking with other NGOs at the national scale helps to maintain a high profile for critical responses to national security legislation, whilst a range of formal and informal networks with NGOs who monitor EU-scale security directives circulate relevant information amongst NGOs. At the same time, the representational practices of anti-immigration NGOs have been shown to engage in a defensive localism, which represents the international scale as a source of

---

37 A. Rudiger, *Prisoners of Terrorism? The Impact of Anti-Terrorism Measures on Refugees and Asylum Seekers in Britain* (London: Refugee Council 2007). See also <http://www.refugeecouncil.org.uk/policy/>.

security threats due to international conventions that make it impossible to close borders. However, countering this scalar representation can be difficult for pro-asylum and pro-migrant NGOs within a securitized context in which the concept of globalization and open borders is surrounded by fear. The representations of civil rights NGOs form a more direct challenge to defensive localism however, by locating security within internationally-agreed principles operating at the global scale, in contrast to the partiality of national security concerns.

Ultimately, it is worth reiterating the point that while the connection between migration and security is not an inevitable one, engaging with it can be fraught with difficulties. For most NGOs, addressing this linkage is therefore still relatively marginal in their work, as they attempt to address, for example, wider immigration issues around destitution and forced returns, or wider civil liberties issues around identity cards. However, the increasing importance to NGOs of networking as a way of pooling resources and increasing impact can mean that, as with the security professionals described by Bigo, immigration and security can become a *lingua franca*, inadvertently reinforcing a contingent connection that must be handled with care.

Acknowledgements

This chapter is part of an ESRC-funded research project on migration, democracy and security (MIDAS) in the New Security Challenges Programme (project ref. RES22320000137) at <http://www.midas.bham.ac.uk>.

Chapter 8

Asylum, Immigration and the Circulation of Unease at Lunar House

Nick Gill

Introduction

The rise in fear, suspicion and mistrust associated with the war on terror has far reaching implications for minority populations in the UK, with individuals and groups who are seen and identified as different disproportionately affected by heightened security concerns. For example, the number of people categorized as Asians stopped and searched by the police in the UK rose by 75 per cent between 2000 and 2004, compared to a 66 per cent increase among blacks and only a 4 per cent increase among whites.[1] Such impacts have coincided with expanded police powers under the Terrorism Act (2000) according to which, although:

> ... there may be circumstances ... where it is appropriate for officers to take account of a person's ethnic origin in selecting persons to be stopped in response to a specific terrorist threat.[2]

While this so-called 'ethnic profiling' of police activities appears to have little practical benefit (13 per cent of stops and searches under normal police powers result in an arrest compared to just 1.7 per cent of stops and searches on suspicion of terrorism)[3] the statistics reveal the racialized implications of the war on terror.[4]

---

1   Statewatch Analysis, 'UK: Stop and Search: Ethnic Injustice Continues Unabated', *Statewatch Bulletin* 15/1 (2005) pp. 15–16, at <http://www.statewatch.org/news/2005/apr/uk-stop-and-search-2005.pdf>.

2   The police powers granted under the Terrorism Act (2000) are to be used in accordance with the Police and Criminal Evidence Act (1984) Code A: Code of Practice for the Exercise by Police Officers of Statutory Powers of Stop and Search from which the citation is taken (para. 2.25). For a discussion, see D. Moeckli, 'Stop and Search Under the Terrorism Act 2000: A Comment on R (Gillan) v Commissioner of Police for the Metropolis', *Modern Law Review* 70 /4 (2007) 659–70.

3   A. Kundnani, 'Racial Profiling and Anti-Terror Stop and Search', *Institute of Race Relations* (31 January 2006), at <http://www.irr.org.uk/2006/january/ha000025.html>.

4   P. Hubbard, 'Accommodating Otherness: Anti-Asylum Centre Protest and the Maintenance of White Privilege', *Transactions of the Institute of British Geographers* 30/1

As part of the singling out of minority groups, asylum seekers constitute a category that is viewed with particular suspicion and hostility in the UK. Despite increasing concern about 'home grown' terrorism and the internal threat to national security, almost a quarter of all those arrested under anti-terrorism legislation between 2001 and 2005 (232 out of 963 people, or 24 per cent) had previously applied for asylum.[5] Given that the asylum seeking community in the UK constitutes less than 0.1 per cent of the population, this over-representation far exceeds that suffered by most minority groups. Such an outcome is set against a backdrop of sustained hostility towards asylum seekers in the UK throughout large sections of the popular tabloid press: 'The asylum shambles is the sea in which terror most easily swims', the *Daily Mail* attests.[6] Such sentiments have been the subject of a long line of criticisms levied at the printed media by scholars working in the field of forced migration.[7] It has been argued that certain sections of the national press conflate the various types of illegality and migration, overlook the link between international violence, civil wars and asylum seeking, and disseminate inaccurate impressions of the level of welfare benefits accruing to asylum seekers as well as their preferential access to employment and housing markets.[8] Following a number of authors who are critical of the language used to construct the asylum issue in the UK, Coole argues that these deficiencies relate to a broader linguistic framework utilized in the popular press that emphasizes the illegal, untrustworthy nature of asylum seekers on the one hand, and their

---

(2005) 52–65; L. Wacquant, 'From Slavery to Mass Incarceration: Rethinking the "Race Question" in the US', *New Left Review* 13 (2002) 41–60.

5   D. Leppard and J. Ungoed-Thomas, 'Asylum Seekers Form Quarter of All Terror Suspects', *TimesOnline* (15 July 2007) at <http://www.timesonline.co.uk/tol/news/uk/crime/article2076120.ece>.

6   'Britain's Lunatic Asylum Policy', *Daily Mail* (20 January 2003) at <http://www.melaniephillips.com/articles-new/?p=95>.

7   C. Coole, 'A Warm Welcome? Scottish and UK Media Reporting of an Asylum-Seeker Murder', *Media, Culture and Society* 24/6 (2002) 839–52; K. Day and P. White, 'Choice or Circumstance: The UK as the Location of Asylum Applications by Bosnian and Somali Refugees', *GeoJournal* 56/1 (2002) 15–26; N. Finney, 'The Challenge of Reporting Refugees and Asylum Seekers', *Information Centre about Asylum and Refugees* (Bristol: PressWise Trust 2003); N. Finney and V. Robinson, 'Local Press Representation and Contestation of National Discourses on Asylum Seeker Dispersal' (2007) at <www.ccsr.ac.uk/publications/working/2007-01.pdf>.

8   B. Arnot, 'Waiting for the Cameras: Journalism and Humanitarian Crises', *World Refugee Survey* (Washington DC: US Committee for Refugees and Immigrants 1995); N. Chapman, 'Detention of Asylum Seekers in the UK: The Social Work Response', *Social Work Monographs* 176 (Norwich: University of East Anglia 1999); M. Samers, 'The "Underground Economy", Immigration and Economic Development in the European Union: An Agnostic-Skeptic Perspective', *International Journal of Economic Development* 6/2 (2005) 199–272.

copious numbers on the other.[9] In a survey of Scottish newspapers, Mollard found that over twice as many articles depict asylum seekers using words with negative connotations, such as 'scroungers', 'floodgates' and 'bogus', than with positive language.[10] What is more, negative depictions appear to have a significant effect: a MORI poll carried out in 2000 revealed that the average *estimated* level of asylum seeker benefits in the UK was £113 per week, far in excess of the £36.54 per week level at the time of the poll.[11] Similarly, a 2002 newspaper poll recorded that the average estimated share of worldwide asylum seekers coming to the UK was 23 per cent, more than ten times the actual share at the time.[12]

One of the striking features about the association of terrorism and asylum seekers in the UK is its counter-intuitiveness. As Frank Furedi points out, suspicion about a large number of purported risks and safety concerns in modern society persist despite good reasons to be sceptical about their accuracy, and fear of asylum seekers is no exception.[13] Between 1999 and 2006, the number of principle asylum claims received by the UK plummeted from 71,000 to 23,500, representing a 67 per cent reduction.[14] In 2001, as concern over international terrorism mounted, these figures compared to 23 million tourists, business people and students who stayed in the UK and a total of 88 million who passed through the UK's borders during that year, dwarfing the number of asylum applications received.[15] The association between asylum seekers and threats to national security is therefore questionable in the light of the comparatively small magnitude of asylum migration flows.[16] Furthermore, it is reasonable to assume that claiming asylum is an increasingly unattractive route into a country from a terrorist's perspective. Asylum seekers regularly come into

---

9   Coole (note 7). See also G. Philo and M. Berry, *Bad News from Israel* (London: Pluto Press 2004); G. Philo, 'The Mass Production of Ignorance – News and Public Understanding', in C. Paterson and A. Srenberny (eds), *International News* (London: John Libbey 2004).

10   C. Mollard, *Asylum: The Truth Behind The Headlines* (Oxford: Oxfam Poverty Programme 2001).

11   MORI, *Britain Today: Are We a Tolerant Nation? Attitudes to Race, Immigration and Asylum Seeking Today* (London: MORI 2000).

12   MORI, *Attitudes Towards Refugees and Asylum Seekers: A Survey of Public Opinion* (London: MORI 2002).

13   F. Furedi, *Culture of Fear: Risk Taking and the Morality of Low Expectation* (New York and London: Continuum 2002).

14   Home Office, *Asylum Statistics, Second Quarter 2007* (Croydon: National Statistics, Immigration Research and Statistics Service 2007).

15   A. Osborne, I. O'Sullivan, and D. Savage, *Travel Trends 2002: A Report on the International Passenger Survey* (London: HMSO 2002).

16   R. Koslowski, 'Towards an International Regime for Mobility and Security?', in T. Kristof and P. Joakim (eds), *Globalizing Migration Regimes: New Challenges to Transnational Cooperation* (London: Ashgate 2006) 274–88; S. Lavenex, *Security Threat or Human Right? Conflicting Frames in the Eastern Enlargement of the EU Asylum and Immigration Policies* (Florence: European University Institute 2000).

close contact with authorities not only at border control points but also before and after they have passed through the border. If they elicit the suspicion of border control officers, asylum seekers can be immediately detained, without charge or release date, subject to the discretion of unelected, civil immigration personnel.[17] Furthermore, their freedom of movement within the UK is becoming increasingly constrained. Their accommodation has been contingent upon their residence in a particular area since 2000, they often have to check in at local police stations on a regular basis, and adult asylum seekers who cannot show that they have experienced torture have been subject to electronic tagging since 2006. For these reasons, any seriously minded terrorist is unlikely to choose asylum seeking as a way to access the UK if there are alternative routes available.

Another striking feature about the fear of terrorism is its ability to serve as a basis for actions that actually contribute towards the threat of aggression itself. As discussed in the previous chapter, the links between fear, suspicion and unease on the one hand and institutionalized practices of security on the other have been theorized in relation to asylum and migration by Didier Bigo.[18] Bigo suggests that the securitization of migration derives in significant measure from 'our conception of the state as a body or a container for the polity', from the 'fears of politicians about losing their symbolic control over … territorial boundaries', and from 'the *habitus* of the security professionals and their new interests' after the end of the Cold War.[19] He also situates these developments in the wider enabling context of the globalization of surveillance technology and the transnationalization of security practices. Thus, the 'immigrant' is taken to be the manifestation of the collapse of the boundaries between the national and international, and is seen as part of a continuum in which mobility is seen as inherently linked to terrorism, crime, espionage and other illicit practices. As Bigo notes, this security prism has become especially important not just to politicians and security bureaucrats, but to a wide range of actors involved in managing the state, welfare systems, the economy, the media and significant parts of the general public. The unease associated with the immigrant thus becomes a way of inducing appropriate behaviour, consolidating a state represented as being under threat and masking other failures of government.

Furthermore, such practices contribute to insecurity for certain kinds of people and increase the risks they are meant to contain. In the context of minority populations in the UK, we can clearly see the negative effects of intrusive attempts to identify security threats, such as stop and search procedures, when we consider the responses of those communities that experience them first hand. Marginalization and alienation can result from the perceived imperative to identify

---

17   L. Weber, and L. Gelsthorpe, *Deciding to Detain: How Decisions to Detain Asylum Seekers are Made at Ports of Entry* (Cambridge: Institute of Criminology: University of Cambridge 2000).

18   D. Bigo, 'Security and Immigration: Toward a Critique of the Governmentality of Unease', *Alternatives: Global, Local, Political* 27 (2002) 63–92.

19   Ibid., p. 65.

security concerns, and the spectre of 'radicalization' can justify practices that provoke precisely the sort of anti-authoritarianism they are designed to contain.[20] By worrying, wondering and agonizing about security, aggression and hostility can actually be produced: precisely through the practices that 'suspects' have to undergo. In the context of security threats and concerns, it appears that searching for something long enough and hard enough can, given the correct conditions, create that which is sought.

This chapter explores such ideas in relation to one of the myriad sites at which security and the state are produced in the UK: Lunar House in Croydon, the headquarters of what was at the time of my research known as the Immigration and Nationality Directorate (IND).[21] It therefore picks up on some of the conceptual concerns regarding the spatial politics of asylum/immigration and security explored in Chapter 7, but examines them in a different context. It also extends them: while Bigo develops his argument primarily in relation to politicians and higher level security professionals, and Chapter 7 considers the responses of UK NGOs, this chapter focuses on the people charged with running the asylum system at Lunar House.

The chapter draws on research I conducted at Lunar House in 2005 and 2006. The aim was to examine not the ways in which asylum seekers themselves experience unequal relations of power of various kinds (although they clearly do), but the ways in which the workforce, including security guards, asylum caseworkers, interviewers, backroom government employees and immigration system managers are induced to exert power over asylum seekers in ways that lead to their exclusion from national territory.[22] The focus was, therefore, on the employees who conduct and implement control of the UK's borders, rather than those who experience this control. As such, attention is given to the pressures and influences that these actors are under in order to assess the ways in which they experience and manage unease about security.

In total, 37 interviews were conducted alongside participant observation of two high profile asylum advocacy campaigns and detailed textual analysis of promotional materials, policy documents and media coverage. My interviews at Lunar House were conducted with national, management level professionals at the

20   M. Ferrero, 'Radicalisation as a Reaction to Failure: An Economic Model of Islamic Extremism', *Public Choice* 122/1–2 (2005) 199–220; A. Breton *Political Extremism and Rationality* (Cambridge: Cambridge University Press 2002).

21   The Directorate has since been reinvented as the Border and Immigration Agency (BIA), partly due to widespread claims of incompetence, inefficiency and corruption throughout the immigration service, including at Lunar House itself. See, for example, *BBC News Online*, 'Clarke Insists: I Will Not Quit' (25 April 2006) at <http://news.bbc.co.uk/1/hi/uk_politics/4944164.stm>.

22   See also L. Weber, and L. Gelsthorpe (note 17) and F. Düvell, and B. Jordan, 'Immigration Control and the Management of Economic Migration in the United Kingdom: Organisational Culture, Implementation, Enforcement and Identity Processes in Public Services', *Journal of Ethnic and Migration Studies* 29/2 (2003) 299–336.

IND, union members working at Lunar House, campaigners working to improve the conditions within Lunar House, and users. They were complemented by drawing upon evidence contained within a comprehensive study of the experience of both staff and users, published in 2006 by South London Citizens (SLC), an independent local charity. This charity surveyed over 300 staff and users of Lunar House, and received 30 written submissions to their report.[23]

A number of high profile media scandals occurred during the period of my research and between September 2005 and June 2006 in particular. Firstly, a senior asylum caseworker was discovered to be abusing his position by offering visas in exchange for sexual favours.[24] Following this, a cleaning firm that was contracted to clean the building was found to be employing asylum seekers illegally, causing consternation in the national press.[25] As these scandals played out, Lunar House featured in a printed newspaper story on average once every four days, including three stories in *The Mirror*, five in *The Telegraph*, eight in *The Observer*, eleven in *The Mail* and twelve in *The Times*. Security staff, interviewers, caseworkers and managers were each profoundly affected by the demands to which they became subject as a result of these stories. The chapter examines the pressures that such media scrutiny created among these employees and suggests that the printed media acts as a key driver of anxiety, fear and suspicion among employees at Lunar House. However, it also indicates the ways in which this episode created opportunities for local NGOs to contest the operation of Lunar House.

Following Stuart Hall et al's seminal research into the way in which media anxiety can prompt state institutions into over-reacting about particular social 'crises', thereby exacerbating the difficulties that are faced, the chapter argues that the security practices that are executed on the basis of media concerns are liable actually to *produce* conditions and procedures that can be degrading, inappropriate and, ultimately, provocative.[26] This point is especially apparent with respect to the spatial layout of the building, which serves both as a security device, and as a provocation. Indeed, the treatment of asylum seekers as a security threat at Lunar House has elicited the very antipathy, despondency and hostility that they are intended to contain. We can, therefore, identify the same self-fulfilling and self-actualizing nature of fear about security at Lunar House that appears to operate through the relationship between minority communities and other kinds of security measures (such as stop and search) in the UK. It begins in the printed media,

---

23   L. Back, B. Farrell and E. Vandermaas, *A Humane Service for Global Citizens* (London: South London Citizens 2005) at <http://www.londoncitizens.org.uk/files/Lunar %20House%20Final_Small.pdf>.

24   *BBC News Online*, 'Inquiry into "Sex for Visa" Claim' (3 January 2006) at <http:// news.bbc.co.uk/2/hi/uk_news/4576618.stm>.

25   *BBC News Online*, 'Illegal Workers Prompt New Probe' (19 May 2006) at <http:// news.bbc.co.uk/2/hi/uk_news/politics/4995764.stm>.

26   S. Hall, C. Critcher, T. Jefferson, J. Clarke and B. Roberts, *Policing the Crisis: Mugging, The State and Law and Order* (London: Macmillan 1978).

translates into uncompromising security procedures and elicits responses from the asylum seekers who use the building in their turn. Discourses of unease thus produce that which they describe, typifying the contradictory and uncomfortable position that modern states occupy between real and imagined threats to national security.

In the next section, the experience of visiting Lunar House is characterized, with particular emphasis on the queuing procedure for which the site has gained notoriety. Following this, the extraordinary sway that the fear of negative media publicity holds over managers, as well as front-line public servants at Lunar House is examined through a consideration of the experiences of SLC,[27] which used the threat of negative media publicity to secure access to carry out an influential analysis of the practices that take place at the site. Then, the security-justified processes and procedures that are undertaken as a result of the constant threat of negative media publicity are outlined and the provocative character of these policies and procedures are examined.

Experiencing Lunar House: Queues

Lunar House is an imposing, 20-storey office block in the centre of Croydon, a bustling London suburb. Together with Apollo House, a neighbouring tower, it housed in 2005 the Immigration and Nationality Directorate (IND) of the Home Office (see Plate 8.1).[28]

The names of the two towers reference the heady optimism of the 'space age', and their grey, austere concrete bulks reflect the architectural style of the late 1960s.[29] Many people make their initial claims for asylum in the offices of Lunar House, meaning that this is often the site at which the government first encounters asylum seekers, and asylum seekers first encounter the state. While significant numbers of asylum seekers apply for asylum at air- and sea-ports around the UK, the majority apply for asylum from within the country. In 2006, for example, 3,580 asylum applications were received at ports compared with 20,030 received in-country.[30] There are only two locations in the UK where asylum seekers can make within-country claims for asylum: at the Asylum Screening Units in Liverpool and at Lunar House. While it is not possible to provide an accurate estimate of the number of asylum applications processed at Lunar House (despite a number of parliamentary questions relating to this issue, it is felt that such information

---

27 <www.southlondoncitizens.org.uk>.

28 Source: <http://www.ukstudentlife.com>.

29 L. Back, 'Remarkable Things: The Scale of Global Sociology', *Goldsmiths Sociology Research Newsletter* 21 (2006).

30 Home Office, *Asylum Statistics 2006*, in T. Heath, R. Jeffries and S. Pearce (eds), *Home Office Statistical Bulletin* (London: National Statistics 2006).

could only be collected at 'disproportionate cost')[31] around two thousand IND employees worked here in 2005, with another four thousand working elsewhere in the Croydon area.[32]

Plate 8.1 Lunar House, Croydon
Source: <www.ukstudentlife.com>.

At that time the administrative system at Lunar House was struggling to accommodate the number of asylum seekers who apply. The facility had become notorious in the national media for the lines of asylum seekers waiting outside the gates, a notoriety that owed itself not only to the visibility of people seeking asylum, but also to the fact that Lunar House could be represented as besieged, incapable of reducing the number of people waiting outside. Queues began to form as early as 5am every day and were divided into two sections. On one side of the building a huge, purpose-built warehouse held the queue of people routinely renewing visas or passports. This queue often reached five hours in length. On the other side of the building, hidden from view behind the concrete bulk of the main office, asylum seekers queued separately in semi-covered areas, often for even

31   See 'Immigration: Lunar House' in *Lords Hansard* 693, col. WA105 (25 June 2007).
32   L. Back et al. (note 23).

longer. Doors closed at 4pm on weekdays, but there have been accounts of asylum seekers still being seen at 9pm and immigration officials themselves being asked to work until midnight in order to clear the backlog.[33]

The difficult conditions faced by asylum seekers while they wait were highlighted in the report of the SLC enquiry. It recorded cold and draughty waiting rooms, a lack of available information (for example on queuing times or immigration procedures), poor provision for families, a lack of available refreshment, poor and inadequate toilet facilities, an incomprehensible complaints process and unsatisfactory fire safety and evacuation procedures.[34] Given these conditions, the length of queues became a recurrent embarrassment, both an expression of a struggling bureaucratic system. The typical representation of Lunar House itself and of the people it was meant to serve is exemplified in the following quote from *The Observer*, a newspaper not usually known for anti-migrant or illiberal positions: 'It is clear from the sea of humanity that descends on Croydon each day that *even 20 storeys of bureaucrats* cannot cope with the workload.'[35]

What is more, given the purported association between asylum seekers and various threats, the queues were seen by the management team as a security risk. These concerns were made explicit by a senior manager at Lunar House: 'From our point of view, we've got a lot of cost constraints so we don't want to have lots of people waiting – the added security and the added buildings etcetera, etcetera.'[36]

Despite government attempts to reduce the queues, waiting times remained persistently lengthy. In 2004, in response to media scrutiny, the layout of the building and the route of the queues were altered in order to promote a faster throughput of asylum seekers and reduce the numbers who were waiting. Interviewers were also given more time away from their desks for breaks, in the hope that they would be able to provide a more 'efficient' service. In an interview in June 2005, however, the then Minister for Immigration was forced to concede that the 'pig pen'-style queues were still there and that the measures designed to ameliorate the long waiting times had not been sufficient.[37]

This exacerbated an already difficult experience for people seeking asylum, after which they faced a difficult interview in which they must present their case.

------

33   National Coalition of Anti-Deportation Campaigns, 'Is there a "Humanitarian Crisis" at Croyden! [sic]', at <http://www.ncadc.org.uk/archives/filed%20newszines/old newszines/newszine30/newszine30.html>.

34   L. Back et al. (note 23).

35   *The Observer*, 'Welcome to Immigration Central. Please Join the Queue: Your Number is 110,001 ... London' (2 March 2003). It is also worth noting that this quote from *The Observer* employs the analogy of the 'sea' of asylum applications. This is in keeping with the widespread use of water-related analogies with respect to asylum seekers, such as flows, floods, deluges, waves and dams, see D. Turton, *Conceptualising Forced Migration*, Refugee Studies Centre Working Paper Series 12 (Oxford: University of Oxford 2003).

36   Interview with member of IND senior management team.

37   File on Four, 'Immigration', broadcast 21 June 2005, transcript at <http://news.bbc.co.uk/nol/shared/bsp/hi/pdfs/21_06_05_asylum.pdf>.

These testimonies are recorded and are used as evidence for the determination of their claim, both at appeal and in the event of deportation. What is more, should an asylum seeker raise the suspicion of caseworkers sufficiently, there are facilities to incarcerate them at Lunar House pending immediate deportation.[38]

Fear of the Printed Press

The asylum system was therefore operating in ways that were stressful to both people seeking asylum and to staff at Lunar House. Under such conditions, negative media attention regarding the offering of visas in exchange for sexual favours and the contracting of illegally employed asylum seekers raised anxiety levels among the management staff as well as among front line employees, such as security guards, interviewers and caseworkers. A new 'director of communications' (a post that had previously not existed) was employed, reflecting the level of concern and exacerbating the tense working conditions of existing staff. SLC also found evidence of the impact of media scrutiny: 'Staff are under so much pressure – not just the targets, but keeping on top of the pressure that's put on them. They're at the front line of national concern and they have to deal with the psychic burden.'[39]

The means by which SLC gained access to Lunar House is also instructive in this regard. The charity was able to harness the *threat* of negative media attention at Lunar House and turn it to their advantage. SLC is a largely voluntary organization, composed of a diverse collection of churches, unions, schools and other civic organizations. The charity aims to improve the lives of marginalized and disadvantaged groups in the South London area. Although Croydon is part of its geographical remit, however, it set itself a difficult task from the outset with respect to Lunar House, adopting as its goal the exposure of the dehumanizing practices that many asylum seekers were experiencing there. There was no reason to expect that the Lunar House management team, which was also the national-level executive managing committee of the IND at the time of SLC's enquiry, would be responsive or receptive to these aspirations.

Initially, the relationship between the management team and SLC was indeed hostile. SLC chose to distribute tea and coffee from a brightly coloured Winnebago to the asylum seekers who were queuing outside Lunar House on a cold morning. This activity attracted media coverage and served to announce the intentions of SLC in loud and obvious terms, with predictably negative reactions from the Lunar House management team. Under constant (now-credible) threat of further publicity, however, the co-operation of the management team was gradually secured. Although the management team would routinely postpone

---

38   HM Chief Inspector of Prisons, *Report on the Unannounced Follow-up Inspections of Two Non-residential Short-term Holding Facilities: Lunar House, Croydon and Electric House, Croydon* (London: HM Inspectorate of Prisons 2006).

39   Personal testimony cited in Back et al. (note 23) p. 63.

meetings, withhold information, miss deadlines, attempt to cancel appointments and leave very long amounts of time between correspondence, SLC repeatedly made use of the threat of staging another eye-catching public action, such as a parade or distributing more tea and coffee at the front of Lunar House. As one SLC organizer, reflecting upon the process of securing the co-operation of the senior IND management team, explains:

> [*After the initial distribution of tea and coffee*] they were taking us very seriously because they saw what we'd done, we'd got quite a bit of media attention. When we heard from [*senior immigration officials*] 'yes, we'll come to discuss working together on the basis that there is no media' we said 'ok, we agree'. And when they hadn't sent us their response ... we decided that we would stage an action, if only to galvanize the support of the voluntary sector. But having let them know, we then got an immediate response back.[40]

This threat underpinned the success of SLC in achieving their stated objectives. Notably, SLC was able to access Lunar House and to negotiate the co-operation of the senior management team. They succeeded in obtaining substantial levels of IND financial resources to meet their own objectives, in the form of a management-level IND employee who was appointed to work two days a week in order to carry out SLC's recommendations. And they succeeded in altering a range of dehumanizing practices at Lunar House, including removal of some of the 'pig pen' style railings, removal of the prohibition of mobile phone use within the building, the introduction of a customer service booth and the re-organization of interview rooms to ensure greater privacy.

Although questions remain about the extent to which SLC were able to alter fundamentally the operation of Lunar House, the fact that they secured co-operation from the management team points towards the sensitivity of the IND, and the government more generally, to the threat of negative publicity and underscores the pervasive fear of media coverage that runs throughout the organization.[41] This sensitivity demonstrates the influence the media can exert through the mechanism of framing – selectively representing certain perceived aspects of reality so as to

---

40   Interview with author.

41   Another SLC volunteer described his frustration at the continuous, circular delegation of responsibility and consistent discourse of inertia and risk-transference that he encountered when dealing with senior IND officials, suggesting that these strategies replaced blunt unresponsiveness when the management team were forced to engage with the charity. As he explains, 'The structures of government themselves seem on the one hand hard and clear cut and on the other hand that hardness seems to evaporate when you touch it. It's almost as if the language of the government and to some degree of the Home Office itself and the Immigration Service has absorbed so much of the language of its opposition'. Interview with author.

promote particular definitions, interpretations, evaluations and treatments.[42] By harnessing the fear of the media's influence that pervades Lunar House, as well as the sensationalism that the printed press can generate, the management team became remarkably accommodating of SLC's demands.[43]

Circulating Security Concerns

Because of the association between asylum seekers and terrorism in the press, and because the press has such a strong influence within Lunar House, it is no surprise that security at Lunar House has been extremely tight. The exterior of the building was patrolled by uniformed security guards. Their job was to police the queues and to crack down on 'asylum agents' offering illegal employment. All entrants to the building were searched and had to remove watches, keys and jewellery in order to pass through airport-style metal detectors upon entering. Entrants' bags were searched and cameras, mobile phones and other electronic equipment were confiscated and stored in locked cabinets at the gates. Only asylum seekers and migrants wanting to renew visas were allowed past the first desk and asylum seekers could not be accompanied unless they were under eighteen years of age or had documentary evidence to show that they were vulnerable. Security of staff was taken extremely seriously. When I interviewed members of the Immigration and Nationality Directorate's senior management team, I was accompanied at all times: first by a security guard in the foyer, then by a secretary in the lift until my hand-over to my interviewee on the top floor of the building.

Three specific security procedures appeared to have a significant impact upon the way Lunar House was experienced by my interviewees. Firstly, the chairs provided at interview are bolted to the floor to prevent them being used as weapons, and no chairs were provided during large parts of the queuing process. Secondly, interviewers were located behind protective plastic screens in order to shield them from personal attack. Thirdly, the entire senior management team was located not in the main offices of Lunar House, but in a separate office five minutes' walk away, to reduce the perceived security risk to the building as well as the team. These measures have implications for the management, users and workforce at Lunar House.

One way in which security policies can inconvenience workers at Lunar House relates to the experience of senior management. Their physical separation from the main Lunar House building has distinct disadvantages. The SLC report highlighted the difficult position of the senior management team in communicating effectively with front-line staff. Contributors to the report raised concerns that

---

42   R. Entman, 'Framing: Toward Clarification of a Fractured Paradigm', *Journal of Communication* 43/4 (1993) 51–8.

43   C. Sparks and J. Tulloch, *Tabloid Tales: Global Debates Over Media Standards* (Lanham: Rowman & Littlefield 2000).

senior management did not understand the needs of front-line staff, or welcome their ideas. With these sorts of perceptions, clear leadership and visible support of caseworkers and interviewers in the main building had become a high priority for the senior management team. The physical distancing between senior management and workers, however, did nothing to meet these objectives. As one of my interviewees outlined, it was more difficult for managers to keep track of the day to day running of the building, including the working atmosphere and the opinions of middle managers, when they were physically separated from it.[44] It was also more difficult for managers to show clear, visible and immediate leadership in the event of disturbances. In a 2004 survey of staff attitudes the Home Office found that only 13 per cent of the workforce thought that the IND senior management team was in touch with staff. The separation of senior management from staff could only exacerbate these difficulties.[45]

Another effect of these security measures was to provoke the users of Lunar House. One of my interviewees had accompanied a number of vulnerable asylum seekers to Lunar House.[46] His frustration at the way in which asylum seekers were expected to endure the queuing conditions was evident:

> I think it's deliberately making it so difficult that fewer and fewer people will even embark on this process. The whole system is wanting to send a message to the countries of origin "Britain doesn't want you. We will make it really tough. Don't come here 'cos we do everything we possibly can to push you back". It's deliberately so that you know that this is going to be really, really tough.

The lack of seating during the waiting period meant that the conditions of queuing became even more arduous. Another of my interviewees gave the following account.

> You had to stand, you weren't allowed to sit in that queue. There were old people, children, all sorts of people, people who had literally just come off the plane, or had come off the back of a lorry: had to stand. There was this pregnant woman next to me, I think she was an African woman, who was I think leaning against something and [an official] came out and he just abused her and told her that she wasn't allowed to lean. It was dreadful. There were pregnant women being herded and being forced to stand around in a way that it's not right for people when they're old and carrying children to be forced to stand in queues, to be treated as though they have no rights at all.[47]

---

44   Interview with member of IND senior management team.
45   IND Staff Leavers Survey, quoted in Back et al. (note 23) p. 67.
46   Interview with asylum advocate and activist.
47   Interview with asylum advocate and activist.

When applicants reached the head of the queue, the fact that chairs were bolted to the floor during the interview procedure, and that interviewers were protected by a plastic screen, also served to aggravate the asylum seekers who were being interviewed. Fixed seats meant that many of them had to lean a long distance forward in order to make their cases for asylum. In this way, the internal layout of the interview room served as a provocation. The protective plastic screen, for example, meant that asylum seekers had to recount their experiences, often involving intimate or disturbing details, in a loud voice in a public room.

When security practices and procedures are themselves provocative, a degree of escalation can set in, giving rise to what Furedi refers to as a culture of fear.[48] The fact that chairs were fastened into position, and that a protective plastic screen was seen as a necessary measure to protect interviewers, was taken to constitute a strong statement about applicants' characters, leading to resentment and hostility among the users of Lunar House. As a local community spokesperson recognized: the security policies 'create a culture of suspicion which makes asylum seekers feel hostile because they are being treated as though they are not good people'.[49]

For this reason, he saw a direct link between the security measures and the risk of security incidents. He suggested that the protective screen was capable of precipitating aggression among asylum seekers who had been waiting all day to be interviewed:

> If you're a member of staff and you've had five people have a go at you in one morning and really get aggressive, you need that screen. But why did those people get aggressive in the first place? Because the screen was there! They couldn't speak properly! The seats are so far away their personal business everybody can hear![50]

A third effect of polices and procedures justified in terms of security has to do with their influence over the front-line workforce within the building, which includes interviewers, security guards and caseworkers. Security procedures influence these workers directly and indirectly, either through their immediate effects upon the workforce or through the reactions that they engender within the asylum seeking population, which impact in turn upon the workforce.

Directly, there is evidence that employees are generally not happy in their jobs: between 2003 and 2005, 50 per cent of IND workers left within the first two years of their employment.[51] While there is some evidence that this might be the result of a stressful working environment (one third of Home Office employees say that they experience stress 'often', 'very often' or 'always' in comparison to under 14

---

48   Furedi (note 13).
49   Interview with author.
50   Ibid. As spoken.
51   IND Staff Leavers Survey 2005, cited in Back et al. (note 23).

per cent in the UK economy as a whole),[52] SLC also noted that the imperative to enact security procedures is a source of regret and anxiety to staff. As one staff member stated: 'I feel anxious, frustrated and demotivated … I am disappointed in myself because I end up acting in an uncaring, unsupportive way when dealing with customers.'[53]

Another interviewee, an asylum seeker support, worker detailed the case of a security guard she had met whilst queuing. The guard had explained that the way he was treating asylum seekers at Lunar House was a source of shame and disappointment to him. These sentiments were so extreme that he was planning to resign from his post the following week in order to pursue work elsewhere. While higher-tier security professionals and politicians may exploit unease about mobile populations for bureaucratic or personal interests, such accounts indicate that the experience of security is considerably less functional for those charged with policy implementation and the everyday management of institutions.

Bureaucratic inefficiencies combined with security procedures can produce other effects that impinge on the subjectivity of people involved in the system, reduce its effectiveness and obstruct claimants' chances of being granted asylum. Another of my research participants had accompanied a vulnerable asylum seeker to her interview and, after waiting with her for seven hours in uncomfortable queuing conditions, had taken a confrontational approach to the interview and attempted to use the threat of newspaper coverage to secure the co-operation of the interviewer.[54] In response, the interviewer had obfuscated important information from my research participant, refused to divulge details of the claimant's case or engage in any discussion of the legal situation of the claimant. The interviewer also refused to accept new information about the claimant's case, concealed her own identity badge from both the claimant and my research participant, and attempted to confiscate the attendance receipt the claimant had been given upon arrival.[55]

This episode illustrates the ways in which anxieties shaped by security fears, bureaucratic dysfunction and the media circulate through individual human interactions and encounters that can have profound consequences. Here it was the person accompanying the asylum seeker who was provoked, and invoked anxieties about media attention in an attempt to influence the IND interviewer, who in turn responded by withdrawing cooperation and contravening protocol (by withholding her identity). While on some occasions the threat of media attention has served to obtain cooperation by powerholders at Lunar House, on others the attempt to exploit unease has led to an escalation of anxiety, suspicion and hostility. Unease thus circulates with unpredictable effects.

---

52 Home Office Staff Survey 2004, cited in ibid.; S. Webster P. Buckley and I. Rose, *Psychosocial Working Conditions in Britain in 2007* (Health and Safety Executive 2007) at <http://www.hse.gov.uk/statistics/pdf/pwc2007.pdf>.

53 Staff Testimony in Back et al. (note 23) p. 63.

54 Interview with asylum advocate and activist.

55 The attendance receipt constituted the only evidence that the interview had occurred.

Conclusion

This chapter has suggested that a number of critical insights can be derived from place-based research into the constitution of security and the state around the management of asylum and immigration which extend the ideas developed by Bigo. In particular, attention to the everyday detail of how policies are implemented and their effects on the people charged with implementing them, as well as the people such policies seek to control, reveals the provisional and problematic nature of the state with a degree of depth and sensitivity to context. It also helps to reveal how contradictions in policy implementation provide opportunities for committed and creative action to challenge existing frameworks, but also the limits and problems of this process. With this in mind, a number of further reflections can be outlined.

The case of Lunar House is illustrative of a range of negative ramifications of the war on terror. Firstly, it is clear from the sensitivity of the management team at Lunar House to media scrutiny that the governmentality of unease has been a powerful influence on policies and procedures at Lunar House. Much of this stems from the extent to which managers and front line staff have felt themselves to be exposed to the often exaggerated demands and scrutiny of the press. Employers have consequently felt obliged to conduct themselves in ways that appear concordant with the imperative to maintain security, which can be detrimental to their own working environments, the work that they carry out, and the people whose fate rests upon their services.

Secondly, by implementing policies that are at once preventative and provocative, the security practices at Lunar House illustrate a particular paradox of concern about security itself: the self-fulfilling and self-actualizing quality of unease. Procedures that treat individuals as potential security threats invite hostility and can create precisely the sorts of reactions they are intended to contain. In these ways, security procedures can undermine security itself by alienating or humiliating those who are subject to them. This effect remains the case regardless of the degree to which purported security threats in fact exist. Indeed, this effect challenges any neat distinction between security threats and responses by highlighting the mutually re-enforcing relationship between them.

It can be suggested that the self-actualizing quality of security in evidence at Lunar House replays on a variety of scales and in a variety of contexts within the logics of the war on terror more broadly. While Lunar House illustrates this at a micro-level, similar processes are evident in the alienation of communities in response to ethnic profiling and stop and search procedures. Moreover, at the international level, whole countries and cultures can react with suspicion, distrust and violence in response to perceived aggression that is predicated upon the search for terrorists and the eradication of security risks, as the unrest in both Afghanistan and Iraq illustrate. In this light, the self-fulfilling potential of unease about security is a phenomenon that characterizes, typifies and problematizes the war on terror across a wide range of situations.

Thirdly, one of the most worrying aspects of the self-actualizing property of security concerns is the fact that, once responses *are* actualized and elicited from those who are treated/produced as suspects, these responses can *then* be taken as confirmatory evidence of the need for security measures in the first instance. This effect can complete the circle between imagined security risks and heightened security procedures, verifying a proposition that was erroneous at the outset. Though the management of asylum and immigration in the UK has been the subject of intense political attention and bureaucratic reform, the persistence of the idea of certain kinds of mobility as inherently threatening to the sovereignty of the territorial nation state means that the potential remains for future crises framed in terms of security.

Chapter 9

Garden Terrorists and the War on Weeds: Interrogating New Zealand's Biosecurity Regime

Kezia Barker

"Des fleurs
ne portent pas les papiers."
Flowers have points of origin,
but not nationality.

Countries claim flowers,
but flowers ignore them, crossing
borders in the night.
There are no nations,

just different climates
and different degrees of tolerance.
Everything grows according
to circumstance.[1]

New Zealand's borders are constantly under threat, from illegal drugs, plant pests, illegal immigrants ... New Zealand customs ... are all that stand in their way.[2]

Introduction

Invasion, threat, terrorism, security. These terms, all too readily associated with contemporary political concerns over human populations, also have meaning

---

1 G. Rigby, 'Nationalities', *50 Botanical Travellers* (Tyne and Wear: Portcullis Press 1990) p. 46.

2 *Border Patrol*, transcribed 27 June 2005. See TVNZ website at <http://tvnz.co.nz/ view/page/536641/881291>, where *Border Patrol* is described as: 'A reality series about the Customs agents who aim to prevent pest plants, animals, drugs and pornography filtering through the country's borders. The series follows officers of the Customs Investigation Unit and Ministry of Agriculture and Fisheries (MAF's) Biosecurity Enforcement Group as they protect New Zealand from harm. What they find is mind boggling!'.

with the prefix biological: biological invasion, biological threat, bioterrorism and biosecurity. Biosecurity, the control of biological threats, is increasingly being linked to anthropo-security, the control of threats emanating from human behaviour such as terrorism and crime. These links are merging human and non-human populations as the justification for security, the means to enact security and the threat to that security. This merging is evident in the discursive basis for biosecurity, as the language and symbolism of war, terror and security is imported into the realm of biotic associations.

To explore these discursive associations we begin at the site of the Wanganui Bloomin' Artz Garden Show, New Zealand, 2005. This show involved a program of talks with gardening experts and celebrities, entertainment and guided tours of notable gardens, and numerous stalls selling plants, gardening equipment, and arty bits and pieces. Within a large marquee opposite a stall selling portable barbecues, the Horizons Regional Council Plant Biosecurity Team rented a pitch for their weed awareness stall. The stall utilized a war and terrorist theme, drawn together by the banner '*Garden Terrorists: Join the War Against Weeds*'.

**Plate 9.1 The war and terrorism theme at the weed awareness stall,
              Wanganui Bloomin Artz Festival, 2005**
Source: Photograph by Kezia Barker.

An identity parade of pest plants was displayed on straw bales lining the stall with a prize for the highest number of correctly identified specimens. A camouflage net was draped over the top of display walls, which were plastered with posters, and a flashing light added to the atmosphere. A further banner urged the unassuming passerby, perhaps laden down with now suspect exotic garden plants, to '*Join the homefront and help us stop the invasion!*' The pest plant officer in charge of the stall toyed with the idea of wearing camouflage facepaint, but settled for camouflage trousers and a Weedbuster t-shirt.

This example shows how discourse is intrinsically connected to language, institutions, practices and emotional realms, with representation bound up in performance and embodied experience. Biosecurity thus involves material–embodied links between security practices targeting human and non-human populations, as practical measures for the control of non-human elements of nature also require the disciplining of human movement and practices.[3] Emotional links can also be discerned in the similar kinds of fear and dread mobilized by both biosecurity on the one hand and terrorist and counter-terrorist activities on the other.[4] Thus, while the linguistic aspects of biosecurity discourse form the main entry point for the chapter, these are historicized and placed in institutional context in a variety of ways, and linked to other dimensions.

In particular, in this chapter I consider this 'Garden Terrorist' banner together with a series of further examples to demonstrate how biosecurity tactics have consciously emulated the language and symbolism of counter-terrorism in an effort to incite participation in the control of plant material, producing 'conceptual, affective, and symbolic borders between spheres once thought of as distinctly separate'.[5] Each example illustrates in different ways the fraught boundary making and boundary breaking processes that underwrite the geopolitics of identity and security. I ask: why have these associations become part of the way plant biosecurity in New Zealand is understood? What are their effects or implications?

Biosecurity is a major preoccupation for New Zealand. The term 'biosecurity' was coined and first used in legislation in New Zealand in the early 1990s through a widening focus from the term 'agricultural security'.[6] Biosecurity New Zealand (BNZ) an offshoot of the Ministry of Agriculture and Forestry (MAF) became the national lead agency in 2005, responsible for overseeing the entire biosecurity system. Biosecurity is enacted in practical governance through border controls

---

3   A. Donaldson and D. Wood, 'Surveilling Strange Materialities: Categorisation in the Evolving Geographies of FMD Biosecurity', *Environment and Planning D: Society and Space* 22/3 (2004) 373–91.

4   B. Braun, 'Biopolitics and the Molecularisation of Life', *Cultural Geographies* 14/1 (2007) 6–28.

5   A. Kaplan 'Homeland Insecurities: Some Reflections on Language and Space', *Radical History Review* 85/3 (2003) 82–93.

6   Parliament of New Zealand, *Biosecurity Act* (1993) at <http://www.legislation.govt. nz/browse_vw.asp?content-set=pal_statutes>.

and surveillant technologies, through eradication campaigns, and through the more mundane control and management of existing pests.[7] New Zealand has high proportional expenditure on biosecurity activities of NZ$200 million a year, inter-agency coordination of policy by an independent Biosecurity Council, and a high input of technologies, legislation, institutions, persons and activities that integrate to produce a particularly complex biosecurity regime. This prioritization of biosecurity, together with a national history of different attitudes to native species, makes New Zealand an important focus for research into the discourses and practices of biosecurity.

In 2005 I undertook six months of field work in New Zealand, researching sites that were significant for the biosecurity regime's attempts to control invasive non-native plants emanating from domestic gardens. Gardening is a crucial pathway for non-native plants to move into and around New Zealand, with 75 per cent of the country's estimated 25,000 non-native vascular plants brought in as garden plants.[8] The domestic gardener has therefore become a significant actor upon which contemporary biosecurity policies in New Zealand are focused. In this chapter I draw from participant observation and interviews undertaken with biosecurity personnel and members of the gardening public during the running of 'weed awareness' stalls at community fairs and garden shows. The somewhat mundane example of the garden show, more the haunt of the retired gardener than enemies of the State, provides the resources to consider the theoretical, material and discursive links currently being drawn between the securing of human and non-human populations, or between anthropo-security and biosecurity. In particular, this methodology offers insights into how security practices are implemented, experienced, resisted and renegotiated in everyday settings that cannot be derived from the study of policy texts or policymaking institutions alone.

In discussing discursive links between anthropo- and biosecurity, I complement recent scholarship on biosecurity with a vein of work in cultural and environmental geography interrogating the native/alien construct in conservation theory and practice.[9] The policing of biotic immigration in the interest of original ecological

---

7   S. Hinchliffe and N. Bingham, 'Securing Life – The Emerging Practices of Biosecurity', *Environment and Planning A* 40/7 (2008) 1434–51.

8   W. Green, 'Biosecurity Threats to Indigenous Biodiversity in New Zealand – An Analysis of Key Issues and Further Options' (2000) at <http://www.pce.govt.nz/reports/allreports/biosecurity_threats.pdf>; M. Jay, M. Morad and A. Bell, 'Biosecurity, a Policy Dimension for New Zealand', *Land Use Policy* 20/2 (2003) 121–9.

9   See for example, A. Kendle and J. Rose, 'The Aliens Have Landed! What are the Justifications for "Native Only" Policies in Landscape Plantings?', *Landascape and Urban Planning* 47/1 (2000) 19–31. P. Green, 'Riparian Alien Plants: Towards Ecological Acceptance?', *ECOS* 23/2 (2002) 34–41; M. Harper, 'Transformers, Neophytes and Aliens – Tackling Non Native Invasive Species', *ECOS* 23/ 2 (2002) 27–33; N. Hettinger, 'Exotic Species, Naturalization and Biological Nativism', *Environmental Values* 10/2 (2001) 193–224; J. Peretti, 'Nativism and Nature: Rethinking Biological Invasion', *Environmental Values* 7/2 (1998) 183–92; D. Simberloff, 'Confronting Introduced Species: A Form of

assemblages is critiqued within this literature as an example of 'nature/culture' dualistic thinking, dependent on arbitrary spatial and temporal boundaries and on the classification of humans as separate from nature. What I will draw from this literature to formulate a response to the Garden Terrorist banner moves beyond concerns with classificatory criteria to focus on what has been termed the 'taint of bioxenophobia' within native/alien distinctions.[10] Critics draw conceptual and empirical links between the nativist value system in relation to 'nature' and racist or xenophobic attitudes to humans, and highlight the exclusionary message that 'native good, aliens bad' has for ethnic minorities.[11] In the context of New Zealand's bicultural settler society this is complicated, as the politically significant ethnic minority, the Maori, have spiritual and ancestral links to the native biota. However, a growing literature interprets New Zealand's use of native nature as national identity to be representative of a postcolonial anxiety that attempts to portray bicultural harmony and to naturalize stewardship of the environment by 'all New Zealanders'.[12]

In what follows, I first describe the use of anthropo-security discourse to legitimize a shift in the object of concern for the emerging biosecurity regime during the 1990s. Drawing on interview material, I then look at how aspects of that regime are managed by those tasked with implementing it. I then consider the discursive tactics at work in the merging of anthropo- and biosecurity and the politics of identity that go with them.

Drawing upon interview material as well as analysis of policy texts and programmes, I aim to highlight the fractured, uncertain and ambivalent institutional and practical dimensions to associations between anthropo- and biosecurity. Conservation managers and invasive plant ecologists have argued that social scientists willingly overlook the subtlety and sophistication of conservation science and management in their desire to represent these practices as examples of binary thinking. I respond to this criticism by drawing upon my fieldwork to access the complexity beyond representational practices. By attending to the complex

Xenophobia?' *Biological Invasions* 5/3 (2003) 179–92; T. Smout, 'The Alien Species in Twentieth Century Britain: Constructing a New Vermin', *Landscape Research* 28/1 (2003) 11–20; C. Warren, 'Perspectives on the "Alien" versus "Native" Species Debate: A Critique of Concepts, Language and Practice', *Progress in Human Geography* 31/4 (2007) 427–46, p. 427.

10   Warren (note 11) p. 435.

11   J.L. Wong, 'The "Native" and "Alien" Issue in Relation to Ethnic Minorities', *ECOS* 26/3 (2005) 22–7.

12   See for example, K. Dodds and K. Yusoff, 'Settlement and Unsettlement in Aotearoa/New Zealand and Antarctica', *Polar Record* 41/217 (2005) 141–55; F. Ginn, 'Extension, Subversion, Containment: Econationalism and (Post)colonial Nature in Aotearaoa/New Zealand', *Transactions of the Institute of British Geographers* 33/3 (2008) 335–53; L. Head and P. Muir, 'Suburban Life and the Boundaries of Nature: Resilience and Rupture in Australian Backyard Gardens', *Transactions of the Institute of British Geographers* 31/4 (2006) 505–24.

politics of legitimacy performed by the banner, I consider how plant biosecurity both parallels and intertwines with counter terrorism, and how people contest, ignore and rework that in an ongoing fashion in everyday life. However, while I am somewhat sympathetic to the motivations of many biosecurity practitioners, I argue that the embrace of security within representational practices remains problematic. Overall, I argue that attempts to play on associations with war and terrorism create an unsustainable tension with parallel attempts to claim non-purist scientific objectivity for plant biosecurity. More sinisterly, these associations normalize, trivialize and imply political closure in this fraught domain of human politics. However, the 'irresistible hook' of these emotive comparisons to catch public attention has captivated academic analysis also, to the detriment of rigorous empirical analysis.[13] In the following, I suggest the need for academic attention to look beyond rhetoric towards the implementation of policy and to everyday practice, and to allow for sensitivity and complexity in the analysis of institutional practices.

Anthropo-Security in the Forging of the Biosecurity Regime

The Garden Terrorist banner urging passers-by to '*Join the War Against Weeds*' can be set within a history of emotive claims about the severity of the ecological damage being inflicted by invasive non-native species on New Zealand's native ecology, and increasing governmental interest in biosecurity. In this context, the current use of terror metaphors and anthropo-security comparisons can be traced to debates about 'environmental pest' plants and the policy regime that resulted from this concern, well before 2001.

New Zealand is frequently cited as leading the world with the most comprehensive and integrated biosecurity system, distinguished by its status as an island nation and its strong focus on border protection.[14] The biogeographical isolation that this biosecurity system is attempting to preserve has been both a blessing and a curse. It has contributed to creating unique ecological assemblages with 80 per cent plant endemism, and to producing a biota particularly fragile to invasion. That invasion came in two waves with Maori settlement in the tenth century and with European settlement in the eighteenth century. European settlement was conceived with a distinct nationalistic agenda in the explicit attempt to remake New Zealand from the soil and grass up as the 'Britain of the South'.[15]

---

13   Green (note 11) p. 34.

14   M. Fasham and K. Trumper, *Review of Non-native Species Legislation and Guidance* (Bristol: Department for the Environment, Farming and Rural Affairs 2001); Jay et al. (note 10).

15   A. Clark, *The Invasion of New Zealand by People, Plants and Animals, South Island* (New Brunswick: Rutgers University Press 1949) p. 384; N. Clayton, 'Weeds, People and Contested Places', *Environment and History* 9/3 (2003) 301–31.

The rhetoric of conquest had a biological twist as European species were seen as superior to New Zealand natives, and bound to win in the Darwinian 'survival of the fittest'. This rhetoric encompassed Maori people as much as the animals and plants. Over the course of the twentieth century a shift in the attribution of value from non-native to native species occurred, due to concern over the environmental impact of invasive non-native species, and a realignment of a New Zealand national identity with native nature. Conceptions of how to improve the ecology of the country therefore underwent a complete reversal, but the construction of this as a patriotic act remained.

From the Control of Thistles Act (1836), the first legal enactment regarding the control of a plant species in New Zealand, to the 1980s, biosecurity concerns were firmly agronomic, as policy makers sought to secure a farming system that was also emblematic of a Eurocentric nation-building project.[16] The shift of the focus of biosecurity to 'environmental pests' was due to a number of contributory factors. These included the rise of environmentalism in 1980s, the influence of constructions of Maori values towards native nature, and the introduction of the Biosecurity Act (1993) that broadened the definition of what constituted a pest.[17] Crucially, however, it required the intense mobilization of concern by environmental protagonists both internal and external to the biosecurity regime, and drawing on discourses of terror and anthropo-security to generate a sense of threat was a key tactic. This is apparent in publications and pronouncements produced by particular environmental agencies. One example is the Parliamentary Commissioner for the Environment (PCE) publication *New Zealand under Siege: a Review of the Management of Biosecurity Risks to the Environment*, a report highly critical of the operational weight placed on agriculturally or economically significant pests.[18] Despite its relatively late nature, the report was instrumental in prompting a full review of biosecurity strategy. The preface makes an explicit comparison between human and nonhuman invasion:

> New Zealand is under siege. Potential animal and plant pests are battering our defence systems ... Our lines of defence are becoming more sophisticated but are not providing the level of protection needed to match the nature and extent

---

16   Parliament of New Zealand, *Control of Thistles Act* (1836) at <http://www.rangi.knowledge-basket.co.nz/gpacts/public/text/1836/an/030>; Jay et al. (note 10).

17   Parliament of New Zealand (note 8); K. Worsley, 'Pest Plants and Their Control' (1999) at <http://www.rnzih.org.nz/pages/1999Conference>.

18   Parliamentary Commissioner for the Environment (PCE), *New Zealand Under Siege: A Review of the Management of Biosecurity Risks to the Environment* (Wellington: Parliamentary Commissioner for the Environment 2001); M. Jay and M. Morad, 'The Socioeconomic Dimensions of Biosecurity: The New Zealand Experience', *International Journal of Environmental Studies* 63/3 (2006) 293–302.

of the invasion ... the current system could be likened to a very famous battle in
the skies over England in the summer of 1940.[19]

This heroic language emphasizes the proximity of battle: the enemy has already
encroached *within* national borders. The Ministry of Research, Science and
Technology also argued for greater attention to biosecurity, suggesting that national
security be extended to incorporate biosecurity:

> Nothing is more central to Government functions than national security.
> Identification of threats to sovereignty and economic prosperity and the provision
> of effective defences are at the core of national defence policies. Protection of
> our biological assets must be seen in the same policy framework ... biosecurity
> needs to *become a component of national security*.[20]

Contemporary biosecurity legislation in New Zealand now appears to match
these highly emotive statements of concern. Considerable powers are invested in
operating authorities. Under the Biosecurity Act (1993) the Minister of Biosecurity
has powers to declare a 'Biosecurity Emergency' in the event of a major incursion.
This allows full access to property and possessions, powers to seize and destroy,
and powers to prevent all movement in and out of risk zones. The high level of
surveillant visibility of plants thus prefigures the current surveillance of human
populations as part of the war on terror. All organic entities within national space
must be fully accounted for: MAF's 'Biosecurity Index' purportedly lists all
plant species that were in New Zealand on July 1997; the Biosecurity Act (1993)
contains a 'black list' of thousands of species termed 'Unwanted Organisms' that
are not eligible for import into New Zealand under any circumstances; any 'new
organism' must undergo a strict importation assessment to ascertain its capacity
to behave as a good, benign citizen; and all plants that receive permission to enter
New Zealand must be accompanied by Import Health Certificates, 'plant passports'
that prove they have undergone the stipulated phytosanitary regulations.

Internal pest management has been placed at the forefront of New Zealand's
plant biosecurity efforts in recent years, comprising over half of the total expenditure
on biosecurity activities.[21] This is due to the strict controls effectively halting new
plant imports, and the huge numbers of non-native plants imported before the
current ban, which have the potential to become invasive. Because it would be both
impractical and politically difficult for the New Zealand state to implement their
powers of access to the private sphere, the extension of biosecurity governance to
the domestic garden makes the success of biosecurity policies reliant on the actions

---

19   PCE (note 20) p. 2.

20   D. Penman, *Managing a Leaky Border: Towards a Biosecurity Research Strategy*
(Wellington: Ministry of Research, Science and Technology 1998) p. 1. Emphasis added.

21   Biosecurity Council, *Tiakina Aotearoa – Protect New Zealand* (Wellington:
Biosecurity Council 2003) p. 52.

of the public. However, the mobilization of institutional concern for environmental pest plants and the accompanying stringent legislation does not operate in an ideal scenario of perfect public compliance, and generating this public concern is not easy. In the following section I highlight the relationship between education and enforcement in the production of a biosecurity-compliant and concerned public. I consider the particular difficulties associated with communicating the environmental risk of pest plants, within which the tactic of drawing comparisons between biosecurity and anthropo-security is justified.

Frontlines of the War Against Weeds: Mobilizing the Homefront

The 'homefront' of the 'war against weeds' is situated in the nation's private gardens. However, the traditional difficulties of regulatory extension to the private sphere are compounded by the risk management focus of biosecurity legislation, and the slow temporalities of plants themselves also seem to wilfully contradict the supposed climate of immanent threat.

The Biosecurity Act (1993) provides a structure for internal pest management, laying out the methodology for the development of National and Regional Pest Management Strategies. It is through these policy instruments that New Zealand's 16 regional councils, with pest management responsibilities on public and private land, devise and differentiate their responses to particular pests.[22] The problematization of garden plants centres on everyday practices played out in the traditionally private sphere of the domestic garden. For regional councils, ways of accessing and influencing practices in the private realm include direct enforcement and diffuse attempts to meld behaviour through public communication and education techniques. These practices are intended to operate in conjunction to produce a biosecurity aware and compliant population, as Carolyn Lewis, the Chairperson of the Biosecurity Institute, National Coordinator of Weedbusters and Waikato Pest Plant Officer, explained:

> What you're trying to do with the newspapers is knock off a few people who are already willing to read about it. Then with your shows you're knocking off some people. Then with Weedbuster groups you're knocking off others, and it gets smaller and smaller, and more labour intensive as you go up. Finally at the top you've got a few people that you may have to serve a legal notice on.[23]

So while considerable powers of enforcement are available to regional councils, with which they can penetrate the private sphere, attaining public support is essential to successfully respond to the threat of environmental pest plants through

22 See K. Barker, 'Flexible Boundaries in Biosecurity: Accommodating Gorse in Aotearoa New Zealand', *Environment and Planning A* 40/7 (2008) 1598–1614.

23 Carolyn Lewis, interview 1:2005.

biosecurity: 'At the end of the day, you can legislate until the cows come home, but until you get that public support, the legislation can help that, but until you get the public support, you might as well forget it.'[24]

This engendering of public support is undertaken through a variety of public education activities, of which the weed awareness stall is one. National consistency for these efforts is being sought through a national weeds awareness campaign, 'Weedbusters'. Launched in October 2003, Weedbusters aims to achieve permanent attitude and behavioural changes for the ecological 'greater good' of both human and non-humans: 'The key task of Weedbusters is to change attitudes and behaviours permanently for the greater good of individuals, their communities and ultimately the wider New Zealand environment.'[25]

The first task of Weedbusters has been to convey the concept that some plants are harmful, the platform of acceptance on which more detailed public education is built. Carolyn Lewis described the difficulties of conveying this concept:

> Everyone thinks that if you dump weeds they just rot down, so trying to convince someone that plants are harmful is actually quite a big ask, because its such a slow thing, its not going to happen over night. They're not going to suddenly take over the world.[26]

The slow temporalities of pest plants are a barrier to conveying their negative environmental impact. This is compounded by the contemporary focus of biosecurity legislation on plants at risk of becoming weedy, *before* they have jumped the garden fence. These preventative strikes on 'risky' non-native plants create tension between institutional and public scales of concern and expertise. Plants visibly acting as weeds in the wider landscape seem more obvious candidates for attention. However, the rationalization of institutional resources means these widespread weeds are not prioritized, as once they have such a stronghold in the country, it is virtually impossible to eradicate them.

A further potential avenue to solicit public engagement is the media. However, the difficulty in generating media attention for weeds, particularly in competition with other conservation and biosecurity issues, was a significant justification for the Weedbusters campaign. While an analysis of MAF's news database from 1995–2000 revealed comparatively low numbers of feature articles on plant pests, Carolyn Lewis put it more bluntly: 'weeds are not sexy'.[27] This is the context in which discourses of security become relevant, as reflected in the tactics of the

---

24   Ibid.

25   Department of Conservation, *Weedbusters Strategic Direction 2003–2008* (Wellington, New Zealand: Science and Technical Centre 2004) p. 4.

26   Carolyn Lewis, interview 1:2005.

27   M. McKenna, 'Snakes and Bugs: Permeable Borders and New Zealand's Bio-Economy', *British Review of New Zealand Studies* 12 (1999/2000); Carolyn Lewis, interview 1:2005.

Garden Terrorist banner and other efforts to generate attention for weeds in the media and concern for weeds amongst publics. In the following section I step from these wider issues of the mobilization of concern back to the Wanganui weed awareness stall, to interrogate the biosecurity regime's repertoire of terror, security, and invasion metaphors, and specifically the construction of non-native pest plants as the threatening vegetative Other.

The Vegetative Other as a Bio-Terror Suspect

As an attempt to raise public concern, the Garden Terrorist banner needs to be seen in the context of growing institutional acknowledgement of the significant role of the public in adequately responding to the issue of invasive pest plants.[28] With this in mind, it is worth reviewing the discursive tactics employed in the banner, which construct and reinforce a sense of strangeness and foreignness around the familiar (but non-native and therefore threatening) garden plant.

While highly-prized, popular and familiar ornamental plants make up only a percentage of those targeted by regional councils, it is particularly difficult to establish and maintain their identity as pests. These plants are often imbued with personal meanings and emotional attachments, so encouraging gardeners to remove these plants requires a significant mobilization of concern through a heightened sense of threat. The Garden Terrorist banner proclaiming a 'War on Weeds' does this by directly associating the suspect non-native plant to a terror suspect. This is however, only one of a number of attempts to generate a sense of threat in association with the vegetative 'Other'.

The posters that brightly adorned the display walls of the Wanganui weed awareness stall highlight boldly any country associations in the common names of pest plants, such as Mexican daisy (*Erigeron karvinskianus*), Argentinean pampas grass (*Cortaderia selloana*) and Chinese privet (*Ligustrum sinense*). Susan Timmins, a Department of Conservation Plant Scientist and significant national figure in plant biosecurity explained: 'If there's a choice over different common names we'll always use a country name to make the point that it doesn't belong here.'[29] These associations have found their way into public discourse. Visitors to the weed awareness stalls extensively utilized country names within common plant names. One visitor proudly announced that she had identified one of the world's worst weeds in her garden: 'Indian something-or-other'.[30] Another visitor discussed the posters profiling different pest plants. Pointing to Chilean rhubarb

---

28 Department of Conservation, *Weed Awareness in New Zealand: Improving Public Awareness of Environmental Weeds* (Wellington: New Zealand: Science and Technical Centre 2003); Department of Conservation, *Weedbusters … Working Together to Protect NZ* (Wellington: Science and Technical Centre 2004).

29 Susan Timmins, interview 2005.

30 Wanganui stall visitor 4, 2005.

(*Gunnera tinctoria*) and Chilean flame creeper (*Tropaeolum speciosum*) she remarked to her friend 'Oh, Chile has a lot to answer for!'[31]

Beyond the direct emphasis on the geographical dis/location of pest plants, other negative connotations can be explicitly produced within common plant names. Wellington Regional Council refers to climbing asparagus (*Asparagus verticillatus*) as 'snake grass' in an effort to make it sound less attractive. Perhaps most disturbingly, on a poster on the Wanganui weed awareness stall highlighting the problems and ecological threat of purple loosestrife (*Lythrum salicaria*), the popular garden plant was described as the 'Purple Peril' in a direct association with the racist characterization of Asian immigration as the 'Yellow Peril'.

The use of military, alien and invader metaphors, and other strongly suggestive language such as 'horrible' and 'nasty,' were described by one pest plant officer as a successful 'shock tactic'.[32] These are attempts to make the mundane world of weeds strange, or perhaps strangely familiar, that is, *uncanny*.[33] The weed of the garden and vegetable patch, made the constitutive outsider in the very act of creating a garden, is perhaps too familiar. The original matter out-of-place therefore needs to be further displaced or alienated through these sensational comparisons.[34] In this manner, these discursive associations are performing the cultural work of reinforcing botanical borders, as much as the plethora of biosecurity policies, practices and technological resources reinforces these borders materially.

These examples speak to the aspect of the native/alien debate focused on the possible 'taint of bioxenophobia' and the wider social effects of this form of exclusionary language. A key question for critics of nativism concerns the potential links between the anti-cosmopolitan attitudes towards people and towards non-native species, or the underlying social and ideological motivations for nativist policies. Critiques of the native/alien construct highlight the conceptual and empirical continuities between ecological nativism and Nazi and Fascist ideas, explicitly interpreting the control of alien species as a form of ethnic cleansing.[35] A typical contribution in this vein comes from Peretti, who describes nativism as a 'purist, xenophobic, and racist way of thinking'.[36] This criticism clearly extends beyond discursive representations to the underlying motivations for nativism, through this association between nativist viewpoints and racist attitudes to people. Simberloff responds to these accusations by empirically investigating, and discounting, the possible link between racism and conservation scientists. He highlights as a counter example that the biggest invasive species program ever undertaken was supported by the Mandela government of South African, though

---

31   Wanganui stall visitor 7, 2005.
32   Craig Davey, interview 1:2005.
33   Kaplan (note 7).
34   M. Douglas, *Purity and Danger: An Analysis of Concepts of Pollution and Taboo* (London: Routledge & Kegan Paul 1966).
35   Warren (note 11).
36   Peretti (note 11) pp. 188–9.

this too might be taken as a form of nation-building at a crucial moment of political transition.[37] In response, nativist protagonists contend that the control of alien species is better likened not to ethnic cleansing but to the prevention of global homogenization and the preservation of cultural diversity, equivalent to keeping McDonalds out of India.[38] This debate shows how accusations of xenophobia are easily made, but in themselves lead to closure, by dis/missing the complexities of practice and therefore missing the biopolitical moment. It also highlights the necessity of clearly distinguishing criticisms of representational tropes from criticisms of invasive species management itself.

Through the interview material generated with biosecurity practitioners I can add empirical detail to this largely theoretical debate, which despite crude frames of analysis makes powerful accusations. This material reveals the reflexive yet ambivalent attitudes of biosecurity practitioners towards the emotive context of these discursive tropes. Before discussing these responses, however, I turn to a further, implicitly connected tactic utilized within biosecurity public communication campaigns to generate public empathy, through the appeal to national natural identity and ecological citizenship responsibilities.

Fearing Them, Imagining 'Us': Discourses of Ecological Citizenship

Attempts to generate a differentiated negative outsider find their obverse in parallel attempts to generate a homogeneous New Zealand identity tied to a vision of a national ecological resource. In the following, I compare the Wanganui 'War on Weeds' themed stall with another weed awareness stall I attended the same year. In contrast to the exclusionary discourses utilized at Wanganui, at the Christchurch 'Gardenz' Show this stall focused less on defining 'Them' and more on consolidating 'Us'.

Plate 9.2 depicts one banner utilized on the stall, which proclaims: 'Together we can stop the spread of weeds.' This banner verbally and visually evokes a sense of shared ecological identity and citizenship responsibility, as biosecurity personnel and willing volunteers smilingly cleanse different habitats of the taint of foreign botanical influence. A sense of ecological citizenship is also knowingly produced through other plant biosecurity public education campaigns, as Jack Craw, the biosecurity manager for Auckland Regional Council, described:

> In Northland we had signs put up saying not 'Get Rid of Your Wild Ginger' because no-one wants to hear that from a bureaucrat ... so we had big signs up saying 'Destroy Wild Ginger Before it Destroys *Our* Forests' [speech emphasis]. So you were doing a patriotic thing by destroying it.[39]

---

37 Simberloff (note 11).
38 Hettinger (note 11).
39 Interview 1:2005.

A link is apparent between patriotism and the necessity of violent expulsion of the botanical Other, as the act of controlling invasive species is portrayed to the individual as a national act. This type of message discursively draws on New Zealand's 'clean, green' image, as the Biosecurity New Zealand website extols: 'Pride in our environment ... and deep affection for our native plants ... have become intrinsic features of the New Zealand cultural identity.' It goes on to urge New Zealanders to: 'Be vigilant and protect those things which quintessentially define us as a nation – which make our country unique and special in the world'.[40] Biosecurity discourses therefore attempt to produce a bounded ecological citizenship identity associated with the nation state for both humans and non-humans.[41] This is used to frame insiders and outsiders, and to link citizenship identity to ecological responsibility. This is not, it must be remembered, a new moment of the mobilization of the botanical in national identity, as the attempt to remake New Zealand in the image of Britain was also formulated through an explicit colonial identity project.

In these ways, the linguistic association between 'native', 'nature' and 'nation' is made real, as the connection between national and natural identity through ecological biosecurity draws on the timeless natural past to legitimate a monolithic identity in the present. Just as Head and Muir associate nativism with settler anxieties about their own belonging, referring to the 'redemptionist narrative of native purism',[42] so can contemporary ecological biosecurity be regarded as affective nation building, the attempt to positively re-categorize the settler society as 'natural' through involvement with the native biota. If the act of protecting indigenous nature is an attempt to render those involved native, then just as acclimatization practices re-made New Zealand 'like home' in the image of Britain, biosecurity represents a new round of 'home-making' activity.

But who in the Christchurch biosecurity banner is really included in the 'we' and incorporated as part of its 'together-ness'? The homogenizing effects of the banner contrast jarringly with the emphasis on the vegetative Other produced by embedding references to foreign countries in the classification of pest plants. The 'Together–We is certainly not 'Argentinean', 'Chilean' or 'Chinese' (pampas grass; flame creeper; privet – does this vegetative-ness get lost?). These country names, now metaphors of illegal spatial mobility, are thrown into relief against the putatively static, timeless, native ecological 'We'. The two weed awareness stalls demonstrate the necessity of the constitutive outsider in supporting any definition of who 'we' are. Discursively citing or inciting the 'foreign' enables the formation of discursive, conceptual and ultimately material/ecological boundaries 'that enclose the nation as home'.[43]

---

40   Biosecurity New Zealand, 'Biosecurity New Zealand' at <http://www.biosecurity. govt.nz>.

41   D. Bell, 'Liberal Environmental Citizenship', *Environmental Politics* 14/2 (2005) 179–194.

42   Head and Muir (note 14) p. 521.

43   Kaplan (note 7) p. 86.

Plate 9.2 'Together we can stop the spread of weeds': A selection of posters at the Gardenz weed awareness stall

Source: Photograph by Kezia Barker.

The unifying 'Together–We' of the banner can be linked to scholarship on New Zealand's postcolonial context that highlights the homogenization of New Zealand identity. In this growing vein of work, nativist conservation and discourses of responsible environmental stewardship have been theorized as attempts to perpetuate an understanding of New Zealand as a benign settler colony – assimilating Maori/Pakeha into the bicultural harmony of 'New Zealander'.[44] The 'Together–We' banner can therefore be situated in a sociopolitical context where competing interest groups have unequal access to claiming knowledge of national goals and identities.[45]

I feel uncomfortable, however, with this easily made suggestion that Maori and non-European immigrant communities in New Zealand are marginalized through ecological biosecurity. It renders these communities too passive. Maori environmental values, imported into the public arena through the 'Maori renaissance', have been cited as a factor prompting a rise in interest and concern for native nature, a crucial basis in the transformation of agricultural security to biosecurity in the 1980s.[46] The necessity to consult Iwi is an aspect of the border assessment process for new organisms, and one of the few new plants to receive approval for import was sponsored by a Maori consortium. A further example of the multicultural complexity of biosecurity politics is the work of the Chinese Conservation Trust in New Zealand. This group, emanating from within the Chinese community, aims to portray positive images of Chinese environmental participation in New Zealand. Activities organized for members have included visits from New Zealand custom officers and the more popular biosecurity ambassadors, beagle sniffer dogs. I am not arguing that nativist conservation and its public representation is necessarily benign. The use of monolithic exclusionary and inclusionary discourses is problematic, but in more complex ways than allowed for in existing critiques of nativism. Further research is needed to consider these interplays and understand the ways in which non-European groups interpret or influence biosecurity ideals.

---

44   K. Dodds and K. Yusoff (note 14).

45   M. Mackenna 1999/2000 (note 29).

46   Prior to the 1960s Maori people lived predominantly within a tribal structure, and worked as rural laborers. With the increasing mechanism of agriculture after WWII, many moved into cities in the 60s and 70s in search of work. While this led to a breakdown in tribal society and language, Maori people became a visible part of the workforce and the heart of trade unions. This fed into the 'Maori Renaissance', with Maori influencing trade unions, anti-Vietnam protests, arts and culture, and participating in university and political life. The subsequent emergence of 'middle class' Maoris led to the infusion of Maori environmental knowledge and values into New Zealand society and culture, significantly including the valuing of native plants.

Terror Insecurities: Scientific Objectivity, Far Right Politics and the Reappearing Banner

I have argued that the language of terror, security and war, has been drawn upon to generate institutional and public concern for plant biosecurity, in the context of slow plant temporalities, an institutional focus on preventative action, and the perceived necessity for public participation. This has been aligned to the production of national identity and a sense of citizenship responsibility attached to native nature, with these inclusionary and exclusionary discourses operating together to demarcate conceptual and physical borders. In the following, I discuss a moment of institutional reflexivity connected to the Garden Terrorist banner, and suggest that this is an example of the tension between sensationalism and a desire to be seen as scientific and rational, a tension heightened by moments of true proximity to fundamentalism and fascist politics.

Carolyn Lewis described how she nearly used the same banner at an earlier garden show: 'I was at a show just after September 11th, and I got out the banner and thought, "Oh, I'll use this one" [pause] And I looked at it and thought, "No I can't!"'[47]

This retreat from terrorist comparisons when it was too highly emotive after 11 September 2001 and its re-adoption at the Wanganui Bloomin' Artz Show in 2005 is important to attend to in the context of debates over the ideological underpinning of nativist policies. Carolyn Lewis chose not to display the Garden Terrorist banner in 2001 as the temporal proximity to the terrorist attacks on the World Trade Centre would have made the association too real and inexcusably insensitive. Rather than highlight the seriousness of invasive plants, the vegetative terrorists would have seemed banal in comparison. Equally, the socially sensitive nature of some of the more emotive exclusionary language has been cemented in association with far-right politics in New Zealand, after its adoption in the rhetoric of the New Zealand First leader Winston Peters. This spatial proximity to fascist politics has challenged the unproblematized adoption of this language, and so its use in biosecurity public communication campaigns is increasingly being avoided. The use of the banner in 2005 must partly be connected to the material presence of this discourse in existing material resources. More significantly it suggests that the attraction of the language of anthropo-security is indeed difficult to resist, despite its problematic associations. It must also be negotiated, both in relation to other events to which that language may apply, and with its adoption by political entrepreneurs advancing visions of New Zealand that are far more exclusionary than those of most biosecurity practitioners. When a current event has not thrown the human or inhuman context of this symbolic language into relief, it seems it can safely and uncritically be drawn on to characterize the botanical world.

The use of these dramatic comparisons between biosecurity and anthropo-security has, however, created difficulties for the biosecurity regime's parallel

---

47   Carolyn Lewis, interview 2:2005.

attempt to claim legitimacy for plant biosecurity by emphasizing its 'objective' scientific status. Not only do these emotive comparisons grab public attention, but they also convey a particular message about institutional attitudes. The use of these discursive tropes has contributed towards the assumption amongst academic critics that nativist conservation is driven by purist or xenophobic attitudes, and this assumption was also evident in the comments of some visitors to the Wanganui weed awareness stall.[48] One gardener remarked: 'It gets to the point where everything that's exotic gets declared bad.'[49] Carolyn Lewis described public reaction to the first weed awareness stall in Hamilton in 1998: 'It was just awful, everyone hated us, it was the most horrible show. "Plant Nazis, you're telling me what I can't grow? But it's a beautiful plant!"'[50]

The 'Plant Nazi' or 'Plant Police' label was clearly a concern for plant biosecurity personnel. Wayne Cowan, Plant Biosecurity Manager for Wellington Regional Council, was keen to explain that he did not have a purist attitude towards native species. He stressed that a non-native species became banned only if it was causing environmental damage. Carolyn Lewis also emphasized the environmental justification behind every pest plant controlled or banned through plant biosecurity policies: 'There is *always* an underlying environmental issue. That's what people don't understand, they call us plant Nazis, think we just don't like these plants. That's not the case.'[51]

Garden shows allow face-to-face contact with the gardening public, which is seen by biosecurity personnel to offer the opportunity to challenge this 'Plant Nazi' label by communicating the specificity of each banned plant, and particularly to explain their own motivations for involvement in pest plant control work.[52] These communicative opportunities were utilized to convey the concept that not all non-native plants are to be banned, by discussing the complex process of assessing each proposed pest plant species based on its potential for environmental damage, not simply its status as non-native.[53] This reveals the desire of biosecurity personnel to be seen as pragmatic and reasonable.[54] However, these claims for legitimacy for plant biosecurity measures as 'objective' scientific responses to environmental damage clearly conflicts with the mobilization of public concern through sensationalized comparisons with anthropo-security, and the onerous screening process that plants must go through before they may be admitted to New Zealand still sets a very high threshold for entry into ecological citizenship.

---

48   Green (note 11).
49   Wanganui stall visitor 11, 2005.
50   Carolyn Lewis, interview 1:2005.
51   Ibid., original speech emphasis.
52   Craig Davey, interview 2005.
53   Mike Harre, Auckland Regional Council public liaison officer, pest plant team, interview 2005; K. Barker (note 24).
54   Craig Davey, interview 2005.

Despite the efforts of those involved in the practical governance of biosecurity in New Zealand, it remains an intensely geopolitical exercise.

Conclusions

In this chapter I have tried to add empirical detail to one aspect of the emotive yet largely theoretical debate over the native/alien construct, by considering the wider context of a 12-foot, plastic-coated nylon banner that made an appearance at a garden show in Wanganui, New Zealand in 2005. I have resisted an immediate reaction of dismissing this banner as an example of bioxenophobia, and I hope that what has emerged adds complexity to the debate. I have argued that the use of the discourses of terror and anthropo-security undertakes the cultural work of reinforcing ecological borders, just as the mobilization of native nature as national citizenship identity dissolves internal borders and conflict in the attempted normalization of pro-biosecurity behaviour. I have situated these discursive practices within attempts to generate both institutional and public concern for invasive plants, and I have highlighted institutional reflexivity and a complex and differentiated response to non-native plants, which had become hidden behind the banner's brash symbolism. However, this discussion has shown the banner to be problematic in two crucial respects.

Firstly, the banner represents spatial and discursive closure where attention to biosecurity practices, policies and perception reveals more flexibility, complexity and openness. This challenges biosecurity practitioners to consider whether the attention gained through comparisons with terrorism is a worthwhile sacrifice for the closure and xenophobia they imply. Biosecurity practitioners do not always have these effects in mind when they choose their discursive themes or catchy slogans, but instead are motivated by their concerns to conserve some of the world's most unique ecological assemblages. However, '[t]he choice of words puts into play a history, but also a present or presence, of multiple meanings and connotations'.[55] These attempts to play on associations between biosecurity and anthropo-security without the harder edges begin to break down when the spatial and temporal proximity to this history and presence becomes too close for comfort. But this insight also necessitates consideration of academic methodologies and commitments: it is much easier to make this banner speak of xenophobia, than it is to hold onto the ambivalent, fractured and uncertain context of its use. For example, the suggestions by critics of nativism that a 'damage criterion' be applied rather than a simple concept of national origins, and that the terms native and alien be dropped in favour of the non-spatialized language of 'pest', 'weed' or 'vermin,' in fact lag behind the adoption of this language and strategy in New Zealand since 1993.[56] However, as the Wanganui banner reveals, this has not

---

55   Kaplan (note 7) p. 86.
56   C. Warren (note 11).

precluded the propensity to draw on binary oppositions within sensationalized descriptive language that inevitably includes and excludes. Given that some kinds of governance of life-forms are necessary, the challenge remains for social scientists to acknowledge the difficulties faced by practitioners and to engage in generating discourses that are not either exclusionary or homogenizing.

Secondly, and more worryingly, this discussion has highlighted the way that war, terrorist and security discourses have become so normalized that they can be transferred and mobilized into what is also presented as an objective, scientific and therefore apolitical context. Carolyn acknowledged that the type of language represented by the Garden Terrorist banner was 'loaded' and had to be used with care:

> Um, we use "alien invaders" and get away with it, which you couldn't use in America, because "alien" means "illegal immigrant". You couldn't use the words "illegal immigrant" here. But it's really tempting, it's a good idea as it gets across the idea of things moving where they shouldn't move.[57]

That the words 'illegal immigrant' might 'get across the idea of things moving where they shouldn't move' reinforces, normalizes, depoliticizes but also trivializes this problematic domain of human politics. The opening voiceover of New Zealand's popular television documentary 'Border Patrol', quoted at the start of this chapter, dramatizes this failure to differentiate between illegal drugs, people and plants. Neither the targets of counter-terrorism or ecological governance, nor the objects that they seek to protect, are as stable, fixed or unproblematic as the practitioners of anthropo-security and biosecurity would have us believe.

---

57   Carolyn Lewis, interview 1:2005.

Chapter 10

'All We Need is NATO'?: Euro-Atlantic Integration and Militarization in Europe

Merje Kuus

Introduction

Discussions of Europe's role in the world often stress the need to strengthen the European Union's (EU) capabilities to project its soft power globally. These discussions refer to the EU in the same breath with NATO, framing both organizations as significant contributors to democracy and security beyond their borders. Formal political pronouncements keep the two institutions strictly separate: EU documents rarely mention NATO and vice versa. The broader rhetorics blend smoothly, however. The speeches and background papers produced by various EU agencies emphasize security and related technical capabilities as a key axis of the Union's external policies, and NATO's proclamations stress societal stability and democracy rather than military threat. For the states that border the two organizations, the term of choice is 'euroatlantic integration'. The Europe and the West toward which these states aspire always include both EU and NATO, and the security in which name they pursue integration is always presented in terms of European values. *The Economist* magazine fittingly remarks that: 'NATO is moving into EU territory; the EU into NATO's'.[1] Through such discursive merger, NATO is increasingly presented not as a military alliance but as a tool for economic development and democracy promotion – almost like an NGO, just more effective. The title quote 'All we need is NATO', cited by one prominent expert as the new mantra of the western security community, encapsulates this narrative.[2]

This chapter examines the discourse that frames NATO as a benevolent institution and a torch-bearer of pacific European norms and values beyond the EU. Empirically, I focus on the ways in which EU and NATO are interwoven in discussions of Europe's neighbourhood, understood here as the states that border

---

1   *The Economist*, 'Berlin Minus: There is No Excuse for the Failure of NATO and the EU to Talk to Each Other' (8 February 2007) at <http://www.economist.com/world/europe/displaystory.cfm?story_id=8669193>.

2   I. Krastev, *NATO in the Age of Populism* (Riga Papers) (Washington DC: The German Marshall Fund of the United States 2006) p. 1.

the two institutions to the east, and its external relations in general.[3] I highlight the trend not only toward greater cooperation between the EU and NATO, but also, and more importantly, toward greater fusion and harmonization of their rhetorics. In so doing, I foreground the normalization of NATO and military institutions more broadly in European political debates. Beyond Europe, the essay contributes to our understanding of militarization – defined here after Cynthia Enloe as a multilayered social process by which military approaches to social problems gain elite and popular acceptance.[4] My focus therefore is on the discourse or logic of explanation that bundles NATO and the EU together as instruments of inclusive, peaceful, and socially progressive power. Within this discourse or consensus, which I call capabilities-talk, EU and NATO appear to go together naturally. Capabilities-talk places security-related institutions, such as NATO, defence ministries and defence think-tanks, at the centre of discussions on EU's external relations. It tacitly habituates European electorates to military activities as an integral part of Europe's 'soft power'.

My argument has two implications for our understanding of the war on terror. First, it contributes to unpacking the ways in which security is diffused into larger areas of social life – a process central to the War on Terror. Only by broadening our analytical focus from the military and the police to other spheres of social life can we bring into relief the processes by which electorates are habituated to the ever more pervasive security agenda. These processes do not offer alternatives to military institutions; they moreover incorporate policy tools like development aid or unimagine the reform of Europe's trade policies. Second, although most geographical research on militarization focuses on the US, Europe is equally important. Analyses of militarization must focus on both sides of the euroatlantic structures. The reframing of NATO is a central part of the geographical imaginary that casts the West as a neutral bystander to violence elsewhere.[5] Precisely at the time when the US appears more aggressively militarized and Europe appears less so by comparison, we must resist such convenient divisions. Instead, we must carefully study the 'soft' security-related consensus in Europe as an integral part

---

3   The term European neighbourhood usually includes countries to the east as well as the south of today's EU. I omit the Mediterranean countries here because these states have less chances of either EU or NATO membership in the foreseeable future.

4   C. Enloe, *The Curious Feminist: Searching for Women in an Age of Empire* (Berkeley, CA: University of California Press 2004) p. 219. The focus here, therefore, is not on military and police capabilities as such, but on the ways in which military institutions are normalized in political arguments. Likewise, my object of analysis is not EU-NATO cooperation as such – whether it is effective or desirable or how it works. The focus rather is on the public legitimation of the cooperation.

5   D. Gregory, 'The Rush to the Intimate': Counterinsurgency and the Cultural Turn in Late Modern War', *Journal of Radical Philosophy* (forthcoming); see also D. Gregory, *The Colonial Present* (Oxford, UK and Malden, MA: Blackwell 2004).

of western military power.[6] We need to examine the War on Terror in places – like the charming capitals of Europe – where it does not seem to be; to analyse closely not only its effects 'over there' but also its legitimation 'in here'.

The rest of the analysis proceeds through three steps. The next section discusses the production of security-related consensuses in the West and the key role of NATO in this process. The subsequent empirical sections illustrate how such consensuses are being made in Europe through the discursive amalgamation of EU and NATO. Drawing from policy analyses as well as EU and NATO documents, I first highlight the shared analytical framework of security and integration in the external relations of the EU and NATO. I then devote a section to how capabilities-talk works in one specific setting – Estonia – to illustrate not just the general EU- and NATO-level rhetoric but also its legitimation in a particular geographical context. The concluding remarks stress that capabilities-talk produces a framework of meaning in which the military apparatus occupies the centre stage. Once political debates are moulded into this framework, they tend to converge on the need to improve capabilities, and capabilities tend to mean the military and the police.

Production of Consensuses

Security is the conceptual axis and glue of capabilities-talk. The framing of complex social issues in terms of threats and the security measures necessary to deal with these threats is a key political dynamic today. This dynamic operates not only in the immediately obvious spheres like foreign and security policy, but also in realms like citizenship, education, and minority rights.[7] Indeed, the gradual shift from a military to a 'human' concept of security since the end of the Cold War has fuelled the proliferation of security threats. Security is now a matter of not only military defence but also social identity, cohesion and loyalty. It becomes detached from defence and is cast in broad terms like 'values' 'cooperation' or integration'. These terms are simultaneously vague and fundamental – they refer to no specific issue yet are presented as crucial for the very survival of a community and everyone in it.[8]

---

6   For further discussion of such dichotomies, albeit in a different context, see S. Žižek, *Welcome to the Desert of the Real* (London, UK: Verso 2002). For European countries' involvement in the war on terror, see J. Sidaway, 'The Dissemination of Banal Geopolitics: Webs of Extremism and Insecurity', *Antipode* 40/1 (2008) 1–8.

7   C. Enloe, *Maneuvers: The International Politics of Militarizing Women's Lives* (Berkeley: University of California Press 2000); D. Cowen, and E. Gilbert (eds), *War, Citizenship, Territory* (New York: Routledge 2008); D. Gregory and A. Pred, *Violent Geographies: Fear, Terror and Political Violence* (New York: Routledge 2006).

8   R. Walker, 'The Subject of Security', in K. Krause and M.C. Williams (eds), *Critical Security Studies: Concepts and Cases* (Minneapolis: University of Minnesota Press 1997) 61–81.

This proliferation of security issues throughout social life is a part and parcel of a similarly multi-sited militarization of social life. Militarization too operates not only at army bases and defence ministries, but throughout civic life.[9] It is manifested in a myriad of everyday details – a can of tomato soup that features military airplanes, camouflage patterns in children's clothing, or the pervasive glorification of military force in popular cinema.[10] Most importantly, militarization operates in ways that make no reference to the military-industrial complex, but rather evoke stability, democracy, and international cooperation.[11] To understand it, geographers need to engage the current period of military conflict without uncritically reifying the role of the state or the military in this process.[12] Empirically, this requires close attention to places far beyond military bases and defence ministries: to everyday practices like aid operations, cultural diplomacy, or youth NGOs – in short, the military-industrial-media-entertainment network that sustains and legitimizes military force.[13] We must take the multilayered character of militarization seriously and carefully reveal the interconnections within that network.

This dynamic of securitization puts great weight on securing not just the 'inside' but also the 'outside' (of a state). It has also resulted in expanding the spatial scope of security measures from the nation to the whole world. As Jennifer Hyndman points out, human security is necessarily expansive in scope as it renders 'governable' all world territory.[14] While (military) defence closes space in an effort to defend it, measures of security lead to opening and globalization.[15] The War on Terror is global in its reach as well as justification.[16] It is based on an integrationist

---

9   M. Kuus, 'Cosmopolitan Militarism?: Spaces of NATO Expansion', *Environment and Planning A* (forthcoming).

10   C. Enloe, *Maneuvers* (note 5).

11   M. Kuus, 'Cosmopolitan Militarism' (note 9).

12   S. Dalby, 'Writing Critical Geopolitics: Campbell, Ó Tuathail, Reynolds and Dissident Skepticism', *Political Geography* 15/6–7 (1996) 655–60, p. 659; Flint, C. 'Terrorism and Counter-Terrorism: Geographic Research Questions and Agendas', *The Professional Geographer* 55/2 (2003) 161–9.

13   J. Der Derian, *Virtuous War: Mapping the Military-Industrial-Media-Entertainment Complex* (Boulder, CO: Westview Press 2001).

14   J. Hyndman, 'Conflict, Citizenship and Human Security: Geographies of Protection', in D. Cowen and E. Gilbert (eds), *War, Citizenship, Territory* (New York: Routledge 2008) 241–57, p. 246.

15   G. Agamben, 'Security and Terror', *Theory & Event* 5/4 (2002); M. Hardt and A. Negri, *Multitude: War and Democracy in the Age of Empire* (New York, NY: Penguin 2004); M. Kuus, 'Cosmopolitan Militarism?' (note 9).

16   See S. Dalby, 'The Pentagon's New Imperial Cartography: Tabloid Realism and the War on Terror', in D. Gregory and A. Pred (eds), *Violent Geographies: Fear, Terror and Political Violence* (New York: Routledge 2006) 295–308; S. Roberts, A. Secor, and M. Sparke, 'Neoliberal Geopolitics', *Antipode* 35/5 (2003) 886–96; L. Bialasiewicz, D. Campbell, E. David, G. Stuart, S. Graham, A. Jeffrey and A. Williams, 'Performing

dynamic in which states are encouraged to participate in global politics under terms favourable to the US.[17] Thus, United States' (US) national security concept emphasizes the need to secure the exterior. The EU's Security Concept as well as the Union's (now failed) Draft Constitution likewise note that the EU's security and defence policies should address areas beyond the EU.[18] This expansive framework in which the whole globe needs to be secured makes connectedness and access (for surveillance) security issues. Threats appear to emanate from non-integration as such. In the words of Thomas Barnett, a key ideologue of this argument: 'in this century, it is disconnectedness that defines danger. Disconnectedness allows bad actors to flourish by keeping entire societies detached from the global community and under their control.' Eradicating disconnectedness, therefore, becomes 'the defining security task of our age' and geographical integration becomes a medium of security policy.[19] In the words of NATO Secretary General De Hoop Scheffer, 'The real issue is this: in dealing with globalized insecurity, it matters less and less where a country sits on the map. What matters more is its mental map – its willingness to engage, together with others, to make a difference.'[20]

This integrationist language loosely mixes concepts like hard and soft power, security and stability, military capabilities and the civil society. These concepts infiltrate into each other, metabolizing into an assemblage in which each element is defined through loose references to all the others. This assemblage, which William Connolly in another context calls a 'resonance machine', is fundamentally amorphous and flexible. It operates through 'energized complexities of mutual imbrication and interinvolvement in which heretofore unconnected or loosely associated elements *fold, bend, blend, emulsify into each other*, forging a qualitative assemblage resistant to classical models of explanation'.[21] When trying to understand how this assemblage works, the task is not to pinpoint causal structures, as those cannot be identified. The task rather is to flesh out the processes of 'mutual imbrication and interinvolvement' that channel political debate in particular directions. Capacities-talk works like Connolly's resonance machine. It is more complex than militaristic rhetoric handed down from the centre of western power; it instead operates through institutional synergies and shared terms of debate in multiple locations. It creates an assemblage in which

---

Security: The Imagined Geographies of Current US Strategy', *Political Geography* 26/4 (2007) 405–22.

17   Bialasiewicz et al. (note 16) p. 405.

18   A. Zumach, 'No Future in Arms', *The Globe and Mail* (21 October 2004).

19   Thomas Barnett, quoted in Bialasiewicz et al. (note 16) p. 405.

20   J. Hoop Scheffer, 'Global NATO: Overdue or Overstretch?', Speech by NATO Secretary General, Jaap de Hoop Scheffer at the SDA Conference, Brussels, Belgium (6 November 2006) at <http://www.nato.int/docu/speech/2006/s061106a.htm>.

21   W. Connolly, 'The Evangelical-Capitalist Resonance Machine', *Political Theory* 33/6 (2005) 869–886. Emphasis in original.

military capabilities are always at the centre of discussion. The constitution of this assemblage in Europe will be my empirical focus below.

Hand in Glove: Euroatlantic Structures

In today's political rhetoric, 'Euroatlantic structures' is in one sense primarily an euphemism for NATO. It does not refer to EU *and* NATO but is rather a synonym for NATO. Yet the term does have a distinct function of its own, and namely, to frame NATO as 'more than' a military alliance – as a key axis of transatlantic cooperation on a deeper level of fundamental values. It is not by chance that all things 'euroatlantic' – as in 'euroatlantic integration' or 'euroatlantic values' – were so popular with candidate states throughout the 1990s. The term 'euroatlantic' seemed to capture the notion that EU and NATO naturally complement each other in one beneficient pacific alliance.

In areas to its east, the EU is traditionally seen as the main actor seeking to enlarge its sphere of influence. The Union actively projects its power beyond its borders through the rhetoric of European norms as well as its economic and social standards (as codified in its law, the *acquis communautaire*). In military terms, the EU is of little relevance compared to the member states. European military capabilities are controlled by the nation-states, who are not keen on pooling these capabilities. The Union does have its own foreign and security policies (the Common Foreign and Security Policy and the European Security and Defence Policy), but these are weak and ineffective. For example, there are plans to build up a 60,000-strong rapid reaction force, but the process is very slow. Europe's power clearly lies in its political, social, and technical standards. One could indeed argue that the Union is so successful in influencing its neighbourhood precisely because it relies not on military instruments but on socioeconomic standards.[22]

The 'big bang' enlargement of 2004 and its follow-up inclusion of Romania and Bulgaria in 2007 put a certain break on enlarging the sphere of EU influence. As the Union's new neighbours (excluding the candidate states) have no realistic chance for membership in the foreseeable future, the EU's task now is to maintain its influence without offering the possibility of membership.[23] The European Neighbourhood Policy (ENP) is a key axis of this effort. It is a policy that coordinates the EU's cooperation with the 15 states that lie beyond its eastern and southern borders. In official terms, the Policy aims at creating a 'ring of friends' – states that have no membership rights in the EU yet align specific policies with the Union in exchange for political and financial assistance. Such assistance includes some free trade as well as political and cultural cooperation in matters like visas,

---

22 *The Economist*, 'Charlemagne: Brussels Rules OK' (22 September 2007).

23 ENP does not include candidate states, which relations with the Union are governed by accession negotiations, or Russia, which has its own bilateral 'strategic relationship' with the EU.

civil society programs, and the transfer of EU know-how. In exchange, the partners must cooperate with the EU on the spheres that the Union considers important for its security. These include transnational crime, communicable disease and a host of issues concerning border management.[24]

Security is the *raison d'etre* of the ENP. The policy was developed concurrently with the European Security Strategy, with which it dovetails neatly.[25] The EU's Commissioner for External Relations and the European neighbourhood Policy Benita Ferrero-Waldner calls the ENP straightforwardly a security policy.[26] The ENP aims to transform societies outside the EU explicitly in order to reduce any security risks that the social problems of these societies (presumably) pose to the EU. All three of the policy's key objectives – prosperity, stability and security – are defined in terms of security. For example, the EU is interested in the prosperity of the neighbours because 'a lack of economic prospects is linked to political unrest, radicalization and is one of the factors pushing people to dangerous illegal migration'.[27] It seeks to promote stability essentially for the same reason – to strengthen state institutions in partner countries as a way to curb crime and illegal migration. In the words of the European security strategy from 2003, '[i]t is in the European interest that countries on our border are well-governed. Neighbours who are engaged in violent conflict, weak states where organized crime flourishes, dysfunctional societies or exploding populations growth on its borders all pose a problem for Europe'.

Military force is nowhere to be found in any materials related to the ENP. However, EU documents and the speeches of key politicians alike speak profusely of the need to enhance the Union's security capabilities. Key speeches stress the need for the EU to 'disperse responsibilities', 'deploy soft power', 'promote democracy' and 'tackle conflict' beyond the Union's borders'.[28] The ENP is presented as a prime example of the EU's 'soft power' – its weapon of mass

---

24   The initial ENP documents spoke of the creation of 'Wider Europe', but this term of subsequently dropped because it was seen as falsely implying the possibility of further enlargement. See J. Kelley, 'New Wine in Old Wineskins: Promoting Political Reforms through the New European Neighbourhood Policy', *Journal of Common Market Studies* 44/1 (2006) 29–55.

25   N. Tocci, 'Does the ENP Respond to the EU's Post-Enlargement Challenges?', *The International Spectator* 2005 (1) 21–32, p. 22.

26   B. Ferrero-Waldner, *Quo vadis Europa?* Speech by European Commissioner for External Relations and Neighbourhood Policy, EPP 'Paneuropa' Group, Strasbourg (14 December 2005).

27   E. Landaburu, 'From Neighborhood to Integration Policy: Are there Concrete Alternatives to Enlargement?', *CEPS Policy Brief* 95 (March 2006) at <http://shop.ceps. be/BookDetail.php?item_id=1305>.

28   *BBC News*, 'EU Should Expand Beyond Europe' (15 November 2007) at <http:// news.bbc.co.uk/1/hi/uk_politics/7095657.stm>.

attraction'.[29] In the words of Commissioner Ferrero-Waldner, 'we need to link intelligently firm action to soft influence, "hard power" to "soft power"'.[30] At the same time, various articulations on EU's security display a marked concern over whether 'soft power' is enough, and whether 'weapons of mass attraction' get things done – in short, whether EU has the proper capabilities.[31] Commissioner Ferrero-Waldner felt obliged to assure her (American) audience in 2005: 'And to those who say we are no military power – more than 500,000 European soldiers are currently keeping the peace and promoting stability around the world.'[32]

This overarching concern with security links the EU to NATO. The latter institution too stresses the need to stabilize and democratize its neighbourhood. It too has enlarged to the erstwhile Eastern Europe twice since the end of the Cold War. Unlike the EU, which consistently maintains that its neighbourhood policy is not a precursor to enlargement, NATO has taken an open door position. Because NATO membership is less focussed on socioeconomic standards than EU membership, the non-members' chances of NATO accession are in fact better than their chances of EU membership. Croatia, Bosnia and Macedonia are often listed as the most probable newcomers. While the membership prospects of Georgia and Ukraine may have diminished, the idea of gaining entry to these organizations has not disappeared. In this context, it is NATO not the EU, that is pushing the geographical boundaries of the Euro-Atlantic structures.[33]

Formal relations between the EU and NATO were established in 2001, but it was only in December 2002 when the first specific cooperation agreement was signed. Since then, the two institutions have collaborated extensively on crisis management. In 2003, they established the EU–NATO capabilities group under an agreement on 'coherent and mutually reinforcing capability requirements'.[34]

---

29   See European Commission, *Non-Paper: Expanding on the Proposals Contained in the Communication to the European Parliament and the Council on 'Strengthening the ENP' – com* (2006) 726 final (4 December 2006); *BBC News* (note 28).

30   Quoted in E. Tulmets, 'Can the Discourse on "Soft Power" Help the EU to Bridge its Capability-Expectations Gap?', *European Political Economy Review* 7 (Summer 2007) 195–226.

31   This concern is articulated through generic references to 'security needs' or 'more robust capabilities', but it pervades the various working papers, speeches, and policy documents on the ENP.

32   B. Ferrero-Waldner, 'Old World, New Order: Europe's Place in the International Architecture of the 21st Century', Speech by the EU Commissioner Ferrero-Waldner at the City University of New York (15 September 2005) at <http://www.europa-eu-un.org/articles/en/article_4029_en.htm>.

33   M. Kuus, 'Cosmopolitan Militarism?' (note 11).

34   P. De Witte, 'Taking EU-NATO Relations Forward', *NATO Review* 3 (2003) <http://www.nato.int/docu/review/2003/issue3/english/art2.html>.

NATO needs such cooperation for political legitimacy and for freeing up its forces for other operations.[35]

This is the so-called 'new NATO'. It proclaims to act not *against* threats but *for* what it calls 'euroatlantic values' – freedom and security, cooperation and solidarity, just and lasting peace, democracy, human rights, and the rule of law. Its space is not closed off in defence against territorial threats, but open in an effort to spread these values. Stretching 'from Vancouver to Vladivostok' and from the Arctic to the Indian Ocean, NATO's network space incorporates Afghanistan and Algeria as well as Canada and Croatia into a distinctly globalist discourse of common security and shared values. The alliance's summits are impeccably choreographed spectacles that project the image of an integrative and inclusive NATO of global reach. In the words of the alliance's Secretary General Jaap de Hoop Scheffer, NATO's paradigm has shifted from deterrence to engagement – from 'being' to 'doing'.[36] Engagement in this rhetoric means the enlargement of NATO membership as well as cooperation with more countries. NATO appears as the politically palatable and socially progressive form of military force. 'These days', remarks *The Economist*, 'NATO looks ever more like a kind of United Nations in military uniform.'[37]

In Europe's eastern neighbourhood, NATO is generally less visible than the EU. Unlike the EU, NATO is not actively involved in civil society initiatives there (although the US is the largest single donor to 'democracy promotion' in Ukraine and Moldova, and third in Belarus after the EU and Germany).[38] Moreover, relations between the EU and NATO are far from smooth, laden with political disagreements over intelligence sharing, costs of missions, and a host of other issues.[39] What the two institutions share is the broad consensus that the Euro-Atlantic West faces 'menacing security threats' and needs to bolster its capabilities to deal with these threats.[40] Both organizations frame NATO as a tool of region-building and integration. Although few 'experts' advocate increased military

---

35 NATO Parliamentary Assembly, *2007 Annual Session: NATO-EU Operational Cooperation* (2007) at <http://www.nato-pa.int/Default.asp?SHORTCUY=1168>.

36 J. Scheffer, 'Managing Global Security and Risk' Speech by NATO Secretary General, Jaap de Hoop Scheffer (7 September 2007) at <http://www.nato.int/docu/speech/2007/s070907a.html>; for an analysis of NATO's image making, see also M. Kuus, '"Love, Peace and Nato": Imperial Subject-Making in Central Europe', *Antipode* 39/2 (2007) 269–90.

37 *The Economist*, 'Berlin Minus' (note 1).

38 In financial terms, US aid to democracy promotion considerably exceeds EU contribution and rivals the combined sum of EU and its largest member states. See K. Raik, 'Promoting Democracy through Civil Society: How to Step up to the EU's Policy Towards the Eastern Neighbourhood', CEPS Working Document No. 237 (February 2006) 16–17.

39 See NATO Parliamentary Assembly (note 35); C. Bennett, 'Building Effective Partnerships', *NATO Review* 3 (2003) at <http://www.nato.int/docu/review/2003/issue3/english/art1.html>.

40 Ibid.

spending, most of them agree that European nations need to develop 'more robust' peacekeeping capabilities – and this means military capabilities.[41] The various security-related documents of both institutions identify various global problems, especially poverty, as the root causes of insecurity, but then speak of the need for enhanced security capabilities. The remark by British Foreign Secretary David Miliband is typical. Discussing the EU's role on global affairs, Miliband said: 'It's frankly embarrassing when European nations – with almost two million men and women under arms – are only able, at a stretch, to deploy around 100,000 at any one time'.[42] The EU's (failed) Draft Constitution from 2004 displays a similar lapse. It first describes socioeconomic processes like underdevelopment and climate change as the root causes of conflict. It then lists terrorism, weapons of mass destruction and 'failed states' as the key threats and concentrates on military instruments to combat these threats.[43] NATO experts (i.e. the individuals who publish in *NATO Review* and related publications) indeed accuse the EU of irresponsible reliance on US military might for its own security needs. In the words of Julian Lindley-French, one such expert, 'so much recent Europeanization has come at the expense of a willingness of Europeans to understand the security implications of globalization and confront them in a realistic and effective manner'. There is, he argues, 'a real danger that a *little* Europe will lead to a *little* NATO, thereby condemning the West and its system of institutionalized and stable power to decline'.[44]

Lindley-French's comments make explicit the terms of debate that frame most discussions. Within capabilities-talk, they come about logically. Consider this bit from the hypothetical report of the EU's second 50 years that *The Economist* produced to mark the EU's (hypothetical) centennial in 2057:

> The other cause for quiet satisfaction has been the EU's foreign policy. In the dangerous second decade of the century, when Vladimir Putin returned for a third term as Russian president and stood poised to invade Ukraine, it was the EU that pushed the Obama administration to threaten massive nuclear retaliation. The Ukraine crisis became a triumph for the EU foreign minister, Carl Bildt,

---

41   F. Cameron and A. Moravcsik, 'Should the European Union be Able to Do Everything that NATO Can?', *NATO Review* 3 (2003) at <http:www.nato.int/docu/review/issue3/english/debate/debate_pr.html>. European members of NATO spend on average just under 2 per cent of their gross national product on their militaries, although there numbers range from over 4 per cent in Greece and Turkey to 1.5 per cent in Germany.

42   *BBC News*, 'EU Should Expand Beyond Europe' (note 28).

43   A. Zumach (note 16).

44   J. Lindley-French, 'Big World, Big Future, Big NATO', *NATO Review* Winter (2005) at <http://www.nato.int/docu/review/2005/issue4/english/opinion.html>.

prompting the decision to go for a further big round of enlargement. It was ironic that, less than a decade later, Russia itself lodged its first formal application for membership.[45]

The article in which the above paragraph appears concludes *The Economist's* special report on the EU's 50th anniversary. It is clearly intended as food for thought regarding future growth and development of the EU. Its most telling aspect is the 'triumph' buried in the middle of the paragraph – that the EU nudged the United States to threaten nuclear war. This is considered as a cause for satisfaction. One can dismiss this as a (intentionally) hyperbolic speculation, but it indicates the normalization of military solutions, in that case the threat of an imminent 'massive' nuclear attack (as if there also exists a non-massive one), as acceptable and indeed praiseworthy.[46]

Formal pronouncements on security integration are only one part of capabilities-talk. This discourse is not simply a high-level rhetoric. Its venues are not as much formal speeches as the more diffuse set of analyses, newsletters, and conference papers that come out of various European and national think tanks. For example, although official ENP-related communications from the European Commission basically omit references to NATO, the military alliance receives much attention in the *European Neighbourhood Watch*, a monthly newsletter on the ENP that is published by the Centre for European Policy Studies, a key Brussels think tank.[47]

This diffuse operation of capabilities-talk also means that the discourse functions through daily political enactments in particular geographical settings. Central Europe offers good examples of such settings. As members of both the EU and NATO, these states actively push for greater involvement of these institutions beyond their borders.[48] The 'Friends of Georgia', group, initiated in 2005 by the three Baltic states, Poland, Ukraine, Romania, and Bulgaria, is one example of

---

45   *The Economist*, 'The European Union at 100: Is the Best Yet to Come?' (17 March 2007).

46   For an analysis of the technostrategic language that normalizes nuclear weapons as legitimate components of politics, see C. Cohn, 'Sex and Death in the Rational World of Defense Intellectuals', *Signs: Journal of Women in Culture and Society* 12/4 (1987) 687–718.

47   For example, see Centre for European Policy Studies, *CEPS European Neighbourhood Watch 37* (2008), or other issues of the *Watch*, all available through the website of the Centre for European Policy Studies at <http://www.ceps.eu>.

48   As NATO is an intergovernmental institution, member states are represented simply on governmental level. In the case of EU, representation includes intergovernmental bodies like Council of Ministers as well as the institutions like the European Parliament and European Commission which involve supranational elements. The European Parliamentary, for example, has its own Cooperation Committees with the ENP states. In terms of the eastern dimension of the policy, these include a Committee for Cooperation with Ukraine, Moldova and Belarus, and another such committee for cooperation with Georgia, Armenia and Azerbaijan. See also M. Kuus, 'Cosmopolitan Militarism?' (note 11).

this.[49] The group seeks to coordinate support for Georgia within the EU and NATO. Its foreign ministers' joint statement from September 2007 argues that the group, together with the United States, see themselves as representing a new centre of gravity within Euro-Atlantic organizations' policies toward Europe's eastern neighbourhood.[50] As such, the group supports Georgia's 'devoted aspirations' to integrate with NATO and the EU.[51] By September 2007, the group had concluded that Georgia already formed a significant element in Euro-Atlantic security and that it was in a position to be offered a Membership Action Plan by NATO.[52] The speeches and interviews by Central European government officials link EU and NATO significantly more than utterances coming from Brussels. They make explicit the linkages between EU and NATO that are often left implicit at the EU- or NATO-level. It was Estonian President Toomas Hendrik Ilves who provided the concluding soundbite on enlargement at the High-Level Expert Conference at NATO's Riga summit. Ilves stressed that NATO must think clearly and pro-actively about enlargement, especially because the EU is failing to do this. Europe, he said, is beset by an enlargement fatigue, 'a complete lack of courage, a complete lack of willingness, a "fortress Europe" mentality' and NATO must not succumb to this. NATO must keep its door open for new entrants.[53] Ilves put the argument in even clearer terms at a security conference in Munich in early 2007. A lengthy quote is warranted to convey his logic of reasoning.

> But if we do not offer the carrot of [EU] membership, are we prepared to see on the borders of the new EU 27, what the EU 15 was not ready to tolerate on its erstwhile borders: unreformed and corrupt governance, illegal immigration, lack of rule of law? More bluntly, are we now ready to allow a zone of lack of prosperity, peace and security on our borders? Perhaps we are.

> There is one alternative of course. NATO. While the trans-Atlantic tie has become weaker in the past 15 years, it might just be that NATO membership represents the carrot that the EU cannot offer ... Perhaps then we should redirect our efforts to bring about reforms in our neighbourhood not through an un-

---

49 V. Socor, 'Friends of Georgia hold strategy session in Lithuania', *Eurasia Daily Monitor* 4/171 (September 2007) at <http://www.jamestown.org/edm/article.php?article_id=2372424>. When Sweden joined in 2007, the group became 'New Friends of Georgia'.
50 Ibid.
51 'Chairman's Summary: Meeting of the Foreign Ministers of the "New Group of Friends of Georgia"', Vilnius (13–14 September 2007); *CEPS Neighbourhood Watch 30* (2007) at <http://www.ceps.be> p. 6.
52 V. Socor (note 49). See the interviews section of the Georgian Ministry of Defence <http://www.mod.gov.ge> for more examples of this line of argument.
53 'Estonian President Ilves: Enlargement Fatigue is EU Problem, Not NATO Problem', *NATO Riga Summit* (December 2006) at <http://www.rigasummit.lv/en/id/pressin/nid/207>.

motivating Neighbourhood programme, but rather through active EU-NATO co-operation and NATO enlargement.[54]

Such statements are significant for the way they fuse EU and NATO, civilian and military cooperation. The rhetoric first evokes 'euroatlantic integration', but the concrete examples of integration concern military institutions and military capabilities. Within it, a country's troop contributions to NATO missions enhance its European credentials. NATO is made an integral part of the European project.

Capabilities-talk in Estonia

Estonia offers illuminating examples of capabilities-talk because the country's position in Euro-Atlantic structures is a prominent issue there. Both EU and NATO are considered essential prerequisites of Estonia's rightful position in world affairs and most political discussions treat the two organizations as two sides of the same coin. Being among the smallest and poorest states in the EU as well as NATO, Estonia is intensely concerned about its image within these organizations. As a result, it is actively building up its capacity to participate in their external relations.

The double accessions in 2004 catalyzed a certain shift in Estonia's foreign policy rhetoric from emphasizing the binary of Europe and Russia to accentuating global issues and global responsibility. Not surprisingly, the ENP is one of Estonia's foreign policy priorities. Estonia's political elites see the policy as the country's niche in the EU (as well as NATO).[55] In particular, they argue that Estonia's historical background and reform experiences make the country an important player in mediating between the EU and its eastern neighbours.[56] Only through such participation can Estonia effectively enhance its standing in the European arena.

This line of argument is even more visible with respect to NATO. After NATO accession, Estonia promptly committed itself to the so-called 8/40 principle, which stipulates that at least 8 per cent of regular army personnel should be in operations

---

54 T.H. Ilves, 'The President of the Republic at the 43rd Munich Conference on Security' (10 February 2007) (Tallinn: Office of the President 2007).

55 U. Paet, *Riigi välispoliitika põhisuunad. Välisminister Urmas Paeti ettekanne Riigikogus* [Main Directions of Foreign Policy. Speech by Foreign Minister Urmas Paet at the Parliament] (7 June 2005) (Tallinn, Estonia: Estonian Ministry of Foreign Affairs); U. Paet, *Riigi Välispoliitika Põhisuunad: Välisminister Urmas Paeti ettekanne Riigikogus* [Main Directions of Foreign Policy. Speech by Foreign Minister Urmas Paet at the Parliament] (7 December 2006) (Tallinn: Estonian Ministry of Foreign Affairs) p. 11.

56 Ojuland, Kristiina. 2004. Address by Foreign Minister Kristiina Ojuland at the Conference 'European Neighbourhood Policy': A Wall or a Bridge? 19 November (July 2007) at <http://www.vm.ee/eng/kat_140/5012.html>.

abroad and 40 per cent of the army should be in principle prepared to join such operations if necessary by 2010.[57] The country actively participates in coaching NATO's partner countries toward closer integration with the alliance.[58] In 2004, Estonia's National Security Concept omitted the notion of territorial defence on the grounds that Estonia had to prepare its troops for NATO missions abroad and the principle of territorial defence hampered these efforts. Estonia has steadfastly supported both the US war in Iraq and the NATO mission in Afghanistan.[59] NATO membership is indeed seen as an even better medium for enhancing the country's international presence because NATO is considered a stronger organization than the EU and its policies more 'real' than the EU's Common Foreign and Security Policy. Ultimately, it is NATO that is believed to make Estonia an international player; to give it capabilities. 'Estonia's security and well-being does not start from the borders of ours state, but from considerably farther', said Foreign Minister Urmas Paet in 2005. 'Membership in EU and NATO has given us the opportunity and means to conduct our foreign policy through this prism.'[60] Estonia should be guided not only by its immediate interests, he stresses a year later, but also by 'the responsibility of the wealthier for the less well-to-do', 'the responsibility of the advanced toward the lagging behind', and 'the responsibility of the member of the club to the candidate'.[61] When Estonia's newly elected President Toomas Ilves visited Georgia on one of his first foreign trips in 2006, he paraphrased John Kennedy's famous 'Ich bin ein Berliner' – 'I am a Berliner' to proclaim 'Me qvartveli var' – 'I am a Georgian'.[62] A few months later, Ilves again stressed 'we must, we absolutely must support and help these barely yet sprung but truly democratic states that suffer from the attacks by authoritarian states like Russia ... To states like Ukraine and Georgia, who have managed to make the choice for democracy, we have to send a clear signal that they have our political support'.[63]

Popular media likewise stress the need for Estonia to actively participate in EU's and NATO's external relations in Europe and beyond. Jüri Luik, an eminent

---

57   U. Paet, *Riigi välispoliitika põhisuunad. Välisminister Urmas Paeti ettekanne Riigikogus* [Main Directions of Foreign Policy. Speech by Foreign Minister Urmas Paet at the Parliament] (6 June 2006) (Tallinn: Estonian Ministry of Foreign Affairs).

58   *Postimees*, 'Eesti jagab Balkanile NATO kogemusi' [Estonia Shares its NATO Experiences with the Balkans] (15 November 2006b).

59   A. Kasekamp and V. Veebel, 'Overcoming Doubts: The Baltic States and the European Security and Defence Policy', in A. Kasekamp (ed.), *The Estonian Foreign Policy Yearbook 2007* (Tallinn: The Estonian Foreign Policy Institute 2007).

60   U. Paet 'Main Directions' (7 June 2005) (note 55). Euro-Atlantic integration is a key sphere of Estonia's foreign aid; for example, with Estonia funding the training of Georgian diplomats and NGO leaders.

61   U. Paet, 'Main Directions' (7 December 2006).

62   *Postimees*, 'President Ilves: ma olen Grusiin' [President Ilves: I Am a Georgian] (23 November 2006).

63   T.H. Ilves, 'Külma sõja järgse ajastu lõpp' [The End of the Post-Cold War Era], *Sirp* (12 October 2007).

foreign policy expert, reminds the readers of Estonia's cultural newspaper *Sirp*, that with EU and NATO membership, the essence of the country's foreign policy will change. 'Estonia will join contemporary postmodern societies that have established a comfortable home in the world. These are also power centers that influence others, we could say, the whole world, with their ideology, economy and sometimes with military force'.[64] Estonians, Luik continues, are 'wholly innocent and inexperienced' in that world, and they need to develop 'global thinking' if they are to be players. Margus Kolga, Deputy Chancellor of the Defence Ministry, likewise stresses the responsibilities of NATO membership. What kind of members do Estonians wish to be, Kolga asks rhetorically: 'egoistic hermits who are interested only in problems affecting them directly; members, who soon after membership forget that the umbrella of collective defence requires responsibility and contribution to the activities of the alliance?'[65] Estonia should rather be a member who actively participates in decision-making. In NATO, says Peeter Kaldre, another foreign policy commentator, 'we cannot explain stepping on rakes with being a transition country. Being in the club means not only that you can use fork and knife in the society – you cannot slurp your soup at home either'.[66] When United States President George Bush visited Estonia in November 2006, his meeting schedule showcased individuals who have 'contributed to sharing Estonia's experiences of EU and NATO integration with other states'.[67] For example, Mart Laar, former two-time prime minister was presented to Bush as an advisor to the government of Georgia. Estonia's President Ilves told his electorate that he did not try to focus his conversations with Bush on Estonia – least of all on any help that Estonia might need from the US. Ilves rather took the conversation to the global issues in which Estonia can contribute to US efforts (e.g. Iraq and Georgia). After Bush's visit, *Postimees* boasted that 'Estonia is no longer a loner in the corner. ... Estonia is no longer a problem, who used to sit at the negotiating

---

64 J. Luik, 'Eesti riik ja ülemaailmne küla' [Estonian State and the Global Village] *Diplomaatia* (October 2003). Reprinted in Eesti Kaitseministeeerium, *Valik riigikaitsealased artikleid* [Estonian Ministry of Defence, 'Selection of Articles on State Defence'] (Tallinn: Estonian Ministry of Defence 2005) 20–22. Luik uses the term 'moodne', which direct translation in this context would be 'modern' as 'in the modern era'. As this would make the phase 'modern and postmodern'. I translate the term as 'contemporary'.

65 M. Kolga, 'Praha tippkohtumisega algas Eestil uus töö' [With the Prague Summit, New Work has Begun for Estonia], *Postimees* (20 November 2002).

66 P. Kaldre, 'Eestist ja NATOst, emotsioonideta' [On Estonia and NATO, Without Emotions] *Maaleht* (12 December 2002). Many of the quotes here come from articles that appeared first in various national newspapers, but were then reprinted in a compendium of educational material on state defence that was put together by the Ministry of Defence. The cover of the compendium features four images: two of military technology and two of soldiers talking to locals clearly in an 'Arab' setting. The cover makes no reference to Estonia.

67 *Postimees*, 'Bush kohtub ka Mart Laariga' [Bush is Going to Meet with Mart Laar Too] (24 November 2006).

table with the US to sadly plead for money, political support or other help. Estonia is [now] a part of the solution'. 'The President of USA met with a partner offering help'. The photo that accompanies the story projects the same message. It shows Bush, his hand extended, eagerly stepping up on a couple of steps of stairs to meet Estonia's Prime Minister Andrus Ansip, who, in a more composed pose, is stepping down toward Bush.[68] In an interview to Estonia's largest daily *Postimees* that same week, Ronald Asmus, Director of the United States German Marshall Fund, likewise frames participation in NATO missions in terms of international responsibility, stature, and progress. He reminds Estonians that their country had prepared for NATO membership in the 'inter-war period', but it now has to address the challenge for the next 40 years: War on Terror.

> One of the main reasons why Estonia joined NATO was to turn its back to the last century. And you have done so successfully. But now it is necessary to turn around and say that Estonia has helped NATO to strengthen security in Afghanistan, Middle East – in countries that many Estonians do not think about. One of the main challenges facing a country like Estonia right now is that the political leaders need to explain to the people why young men and women are on such missions.[69]

These examples show the smooth blending of EU and NATO in discussions of international cooperation. Concepts like cooperation, responsibility, democracy, and military cooperation all are fused into one seamless story. The rhetorical emphasis is on civil society and democracy, but the political process it normalizes is NATO enlargement. In other words, a securitized conception of international cooperation is naturalized through the discourse of capabilities.

Conclusion: NATO Naturalized

This chapter examined the discourse that bundles up EU and NATO as natural partners in efforts to enhance stability and democracy in Europe's neighbourhood. I used the term capabilities-talk to denote a discourse that frames security and related technical capabilities as a key axis for these efforts. In this discourse,

---

68   R. Kagge, 'Eesti riik ole enam nurgas istuja' [Estonian State is No Longer the Loner in the Corner] *Postimees* (29 November 2006). Casting NATO membership as an essential part of Estonia's political subjectivity (as an advanced western state) implies that anyone sceptical of NATO is a loner and a loser. This makes it almost impossible to publicly take a NATO-skeptical position in Estonia. For an exploration of resistance to NATO in Central Europe, see M. Kuus, 'Svejkian Geopolitics: Subversive Obedience in Central Europe', *Geopolitics* 13/2 (2008) 257–77.

69   Quoted in E. Bahovski, 'NATO peab jääma oluliseks' [NATO Must Remain Important] *Postimees* (29 November 2006).

Europe's external relations revolve around its need to secure itself against instability from the outside, these relations thereby naturally overlap with NATO's mandate, and NATO is therefore a natural part of Europe's external relations. This discourse operates not by evoking military threats but through the concepts of international cooperation, stability and democracy. Yet it legitimizes a military alliance as an essential tool for achieving cooperation. It tacitly habituates us to a mode of international cooperation in which Europe's neighbourhood is seen primarily in terms of EU's security and security is seen in terms of NATO. It ultimately legitimizes military capabilities, military institutions and military spending. Its effects are discernible in the widespread acceptance of NATO's existence, enlargement and the use of NATO forces in Afghanistan.[70] Capabilities-talk frames international action in terms of military action – or at least an action by a military alliance – and this at a time when even the so-called security experts agree that the military can provide only a small part of the security of human beings.

In more specific terms, the chapter highlighted three facets in the operation of capabilities-talk. First, this discourse is resistant to causal analysis. EU and NATO documents do not establish any explicit link between the two institutions and frequently do not even mention each other. Capabilities-talk is not a pre-given agenda. Rather, it is the emulsification and blending of loosely associated elements that Connolly identifies as the 'resonance machine'. It does not enforce a uniform understanding of key issues, but it establishes a particular field of articulation. The loose and diffuse definition of security is the glue that holds the discourse together. Thus, although there is a great deal of debate about NATO operations and EU–NATO relations, capabilities-talk provides the shared vocabulary of integration, cooperation, and capabilities that enable particular discussions of EU external relations while disabling others. Although the term capabilities does not always refer to military capabilities (it sometimes denotes diplomatic relations), the profuse references to security capabilities resonate to legitimize military capabilities. They contribute to the sense that Europe is under threat and needs to build up its militaries to deal with that threat. Second, the reverse lending of legitimacy – from the concept of Europe to that of capabilities – is in place as well. Although capabilities-talk is in some sense simply a codeword for military-related resources, its legitimacy is based on linking capabilities to the concepts of Europe and European values. The discourse works precisely because it evokes *non-military* cooperation. NATO appears to counter not just threats but disconnectedness more generally. The concept of Europe is central in this process. Since that concept functions as a signifier for peace in European political discourses, linking NATO to the idea of Europe frames the alliance as a non-military institution. There is an irony here. The post-war European project started as an explicitly non-military project, and yet today, integration with a military alliance bestows European credentials on a country. Third, capabilities talk is not a high-level blueprint externally imposed

---

70   Kuus, 'Cosmopolitan Militarism' (note 9).

on countries. Rather, it is produced in multiple locations to serve a variety of local concerns and aspirations. In Estonia, for example, it functions to alleviate nagging concerns about the country's international clout and standing in the West. The discourse is flexible enough to respond to the particular concerns of many social groups in varied contexts. It is precisely its diffuse character that makes capabilities-talk so effective and yet so difficult to pin down.

Beyond Europe, the chapter highlights the political discourse in which the security establishment, including military institutions as well as the related military-industrial-academic complex, occupy the centre of political debate. This discourse blurs the line between civilian and military-related regional cooperation, political and military institutions, peacetime and wartime. Such blurring is an integral part of the increased powers of the state domestically as well as the global state of imperial war.[71] In Europe, it furthermore promotes the notion that a good way and perhaps the only way to curb American unilateralism is for the EU to build up its own military capacity. For geographic research, these findings underscore the need to investigate the multiple sites, scales, and trajectories of militarization. We need to investigate the civilian rather than the military, the boring rather than the spectacular, the seemingly beneficent rather than the visibly menacing, and the spheres of international cooperation rather than the centers of state power. Militarization starts much before anyone mentions military force.

Acknowledgements

An earlier version of this chapter was presented at the Association of American Geographers conference in Boston on 17 April 2008. I thank Gregory Feldman, Cynthia Enloe, Klaus Dodds, and Alan Ingram for constructive comments on that version. Research for this paper was supported by the Social Sciences and Humanities Research Council of Canada.

---

71   Hardt and Negri (note 15) p. 5; C. Lutz, 'Empire is in the Details', *American Ethnologist* 33/4 (2006) 593–611.

PART 3
Alternative Imaginations

Chapter 11

Satellite Television, the War on Terror and Political Conflict in the Arab World

Lina Khatib

Introduction

The Arab world today is at a political crossroads. Continuing conflict in Iraq, tension in Lebanon, and intra-Palestinian rivalry threaten to destabilize the region. At the same time, foreign intervention in those conflicts shows no sign of decreasing, which is not surprising considering the international nature of politics in the Arab world. From the days of European mandates in the region, to the establishment of the state of Israel and subsequent Arab–Israeli wars, and on to Lebanese Civil War, the Palestinian intifada and the Gulf War, the Arab world has been host to a series of foreign interventions, both political and military. The war on terror has only consolidated and intensified this intervention, so that the notion of geopolitics in the Middle East has come to take on a global, rather than a regional, dimension.

Over the last decade, satellite television has affirmed its place as the primary news medium in the region. The Gulf War was marked by the dominance of CNN, its images of smart weapons and precision bombs colonizing television screens worldwide. The establishment of al-Jazeera in 1996 was the Middle East's first attempt at entering the world of 24-hour news channels. However, although al-Jazeera was a well-respected and relatively well-known channel in the Arab world at the time, it did not enjoy a primary position in people's homes. It was the second Palestinian intifada in 2000 that made al-Jazeera a recognized brand in the region. Al-Jazeera devoted much of its broadcasting time to coverage of the intifada, presenting a clear pro-Palestinian stance towards the issue.[1] Zayani argues that in doing so, al-Jazeera set itself a political role in the Arab world:

> Al Jazeera's intense coverage of the intifada has not only fed Arab fury but also fostered anti-government behavior in the Arab world, making Arab governments vulnerable to charges and open to criticism that they have not sufficiently supported the Palestinians or decisively acted on the Palestinian cause. In this

---

[1] M. Zayani, 'Witnessing the Intifada: Al Jazeera's Coverage of the Palestinian-Israeli Conflict', in M. Zayani (ed.), *The Al Jazeera Phenomenon: Critical Perspectives on New Arab Media* (London: Pluto Press 2005) 171–82.

sense, Al Jazeera places itself as a counter-force to the official indifference towards the plight of the Palestinian people.[2]

At the same time, al-Jazeera's coverage of the intifada marked a significant change in the Arab television landscape: the assertion of the primacy of the image as a means of political communication. Ayish notes that al-Jazeera 'went one step further by showing live footage of clashes in Jerusalem between Palestinian stone throwers and heavily armed Israeli soldiers'.[3] Less than a year later, the events of September 11 consolidated the transformation of Arab satellite television into a visual-saturated medium. They also consolidated the role of Arab satellite television as an active political participant in the region, as opposed to a mere carrier of messages.

The war on terror contributed to the prominence of al-Jazeera in particular, and satellite television in general, in the Arab world. The events of September 11 were constructed as a television landmark, dominating the screens of television channels in the Arab world and beyond. Jean Baudrillard famously said that September 11 attacks were the 'absolute event, the 'mother event', the pure event'.[4] The attacks gave birth to images that have carved a permanent space in the visual memory of people across the globe. The video tapes sent by al-Qaeda to al-Jazeera following the attacks form part of this visual memory. Al-Qaeda's courting of al-Jazeera after September 11 is well documented, giving the station a worldwide notoriety and transforming it into a household name across the globe. Through the war on terror, satellite television in the Arab world grew in presence and impact, establishing itself as one of the most widely consumed media in the region.

In addition, in the decade or so since the war on terror started, satellite television in the Arab world has witnessed much contestation and competition. This decade has seen a proliferation of channels besides al-Jazeera – including non-Arab channels like al-Hurra – that form part of competing international public diplomacy efforts in the Arab world. Within this context, satellite television has moved from a medium seen as providing a space for political dialogue in the Arab world to a challenger to this very space. Satellite television in the region is not only a tool of communication. It is also a symptom and sometimes even a cause of power struggles in the Arab word. These power struggles – especially since 11 September 2001 – are at once national, regional and global. The intra-Lebanese conflict that has consumed the country since the assassination of former Prime Minister Rafic Hariri for example cannot be examined and understood without thorough attention to the roles played by international agents, be they organizations

---

2   Ibid., p. 174.

3   M. Ayish, 'Political Communication on Arab World Television: Evolving Patterns', *Political Communication* 19/2 (2002) 137–54, p. 149.

4   J. Baudrillard, 'L'Esprit du Terrorisme', *The South Atlantic Quarterly* 101/2 (2002) 403–15, p. 403.

(the United Nations and the Arab League) or states (Egypt, Syria, Saudi Arabia, Iran, France, and the United States being the key players).

The various overlapping power struggles in the Arab world play an important role in shaping the visual and political economic television landscape in the region. Satellite television is firmly and actively embedded within this complex structure. It attempts to challenge official political points of view. This is mostly seen in al-Jazeera's coverage of the Iraq war, which challenges the American version of the events. It also engages in processes of political conflict by proxy, becoming a platform for rivalries between Arab countries, clashing 'national' political groups, and international political agents. In doing so, satellite television acts as a mouthpiece for warring political factions. The roles that satellite television plays in the Arab world mean that Arab satellite television has itself become a political actor in the Middle East and beyond. In what follows, I offer a critical assessment of this statement through a historically contextualized examination of satellite television's position within national, regional and international political struggles affecting the Arab world.

The Political Economy of Arab Satellite Television

The emergence of satellite television has created a nexus of power over the Arab television space by competing television stations. The nature of this competition has transformed the landscape of the Arab televisual media from being inherently national, to being regional or pan-Arab.[5] The Arab world's first private satellite channel is MBC (Middle East Broadcasting Center), which was launched in 1991 in London by the son-in-law of the Saudi King Fahd bin Abd Al-Aziz, and relocated to UAE in 2003.[6] A number of other private satellite channels followed suit, such as ORBIT in 1994 and ART in 1995. Meanwhile, terrestrial television channels also started to broadcast on satellite, such as the Lebanese channels LBC (Lebanese Broadcasting Corporation) and Future Television, which launched their satellite channels in 1995. However, it was not until 1996 that a satellite channel fully dedicated to news was launched, and that was al-Jazeera. It started with 6 hours of broadcasting per day, and moved into 24-hour programming in February 1999.[7]

There are economic reasons for this regionalism as privately-owned satellite television stations seek consumers beyond the borders of the country they broadcast from. Economic aims have sometimes contributed to a change in the identity of

---

5   N. Sakr, *Satellite Realms: Transnational Television, Globalization and the Middle East* (London: I.B. Tauris 2002); N. Sakr, *Arab Television Today* (London: I.B. Tauris 2007).

6   M. Kraidy and J. Khalil, 'The Middle East: Transnational Arab Television', in L. Artz and R. Kamalipour (eds), *The Media Globe: Trends in International Mass Media* (Lanham: Rowman & Littlefield 2007) 79–98.

7   Ayish (note 3).

television stations. The Lebanese Broadcasting Corporation (LBC) for example started in 1986 as a channel aimed at the Francophone audience in Lebanon, relying on a relatively high usage of the French language and French programmes in its broadcasts. This language use was partly political, as the channel was set up by factions belonging to the Maronite Lebanese Forces militia, whose target audience was primarily the Christian communities in Lebanon, particularly those who associated themselves culturally with France. In the post-Lebanese Civil War era, LBC has ceased to be an exclusive Lebanese Forces company and become a listed corporation with shares owned by diverse investors. As LBC's popularity in Lebanon grew, its use of the French language decreased as it attempted to appeal to a wider local audience.

However it was the launch of its satellite channel in the mid-1990s that characterized the toning down of the use of the French language in programmes and by presenters and the increase in the Arabic language content of this satellite channel. This proved to be a successful business move, and LBC became the only Lebanese television channel actually generating a profit. In 2003, the Saudi Prince Al-Walid bin Talal bought 49 per cent of the shares in this satellite channel.[8] This part ownership, along with the channel's targeting of wealthy consumers in Saudi Arabia through its programmes, have also meant that its programme content has been geared more towards this audience, with more material being aired that addresses the Saudi Kingdom in direct ways.[9] Other Arab satellite channels have followed a similar path, gearing their programmes towards a wider pan-Arab audience.[10] With the Arab market, rather than the local markets, being the economic focus of most satellite television stations in the Arab world, satellite television has come to unite its viewers by constructing them as consumers. Consumption has become one category that binds Arab audiences together and hides their social antagonisms.[11]

Satellite Television's Promise of Democracy

However, if Arab audiences are 'united' by being consumers, the political factors underlying their consumption of satellite television products present a paradox: at first glance, satellite television seems to offer a potential for the creation of a dialogic political sphere, where audiences get together in sharing a democratic

8 'Satellite Chronicles: November 2003 to April 2004', *Transnational Broadcasting Studies* 12/Spring-Summer (2004) at <http://www.tbsjournal.com/chronicles.htm>.
9 M. Kraidy, 'Reality Television and Politics in the Arab World: Preliminary Observations', *Transnational Broadcasting Studies* 2/1 (2006) 7–28.
10 L. Khatib, 'Language, Nationalism, and Power: A Case Study of Arab Reality TV Show *Star Academy*', *Southern Review: Communication, Politics and Culture* 39/1 (2006) 25–41.
11 M. O'Shaughnessay, 'Box Pop: Popular Television and Hegemony', in A. Goodwin and G. Whannel (eds), *Understanding Television* (London: Routledge 1990) 88–102.

space of expression. But a closer look reveals a more complex story. A decade and a half ago, when satellite television in the Arab world was still in its early stages, this development was hailed as a catalyst of social and political change in the region. Much was subsequently written about the role of satellite television in countering Western narratives about the region, and, with the rise of al-Jazeera, about the potential of this one television station to transform the Arab political sphere.[12] This romanticism is understandable when considered in context of the many constraints on freedom of speech in the Arab world, and the fact that the majority of Arab countries at the time had previously had access only to the television stations owned by the states governing them. Ayish wrote in 1989 that political news on those channels was primarily concerned with reporting leaders' speeches, visits and activities, making their content dull and monolithic.[13] Arab audiences got used to, and learnt to ignore, those state channels that were preoccupied with reporting the whereabouts of leaders while overlooking any sense of mass dissent. Whether it was President Assad of Syria, or Saddam Hussein of Iraq, the image of the leader was a prominent one, continuously relayed to the local audience, and packaged positively: those television stations presented the leaders as benevolent, patriotic and popular. The image of the leader on television was deployed in an effort to 'enforce obedience and induce complicity' in the people, and to produce 'belief in the regime's appropriateness'.[14]

With the exception of the constant presence of the image of leaders on Arab state television during this time, those television channels were notably rhetoric-heavy and light on visual representations. Muhammad Ayish points out that the news formats of state television 'are characterized by serious and formal delivery methods that do not accommodate conversational approaches to news presentations. In this context, newscasters usually appear in an on-camera setting, with little consideration for television as a visual medium of communication'.[15] From 1996 onwards, this situation began to change. The dullness of those local television stations was overshadowed by the comparative sleekness of presentation and content on al-Jazeera. Those audience-attracting tactics drove state-owned television to follow similar patters in using graphics and images.[16] The news as a visual form began to find its way into the living rooms of Arab households.

Al-Jazeera also challenged leader-centric news by relying on a mixture of reports, studio guests and critical analysis addressing a wide range of political and social issues in its news broadcasts. *The Economist* wrote in 2005 that satellite

12   M. El-Nawaway and A. Iskandar, *Al-Jazeera: The Story of the Network that is Rattling Governments and Redefining Modern Journalism* (Boulder: Westview 2003).

13   M. Ayish, 'News film in Jordan Television's Arabic Nightly Newscasts', *Journal of Broadcasting and Electronic Media* 33 (1989) 453–60.

14   L. Weeden, *Ambiguities of Domination: Politics, Rhetoric and Symbols in Contemporary Syria* (Chicago: Chicago University Press 1989) p. 158.

15   Ayish, 'Political Communication on Arab World Television' (note 3) p. 140.

16   *The Economist*, 'The World through their Eyes' (26 February 2005) 23–5.

television is driving state-owned television to follow similar patterns in using more field reporting and less adulatory coverage of Arab leaders.[17] With al-Jazeera favouring heated talk shows over polished representations of political rulers, it was refreshing for Arab audiences to witness what they perceived to be 'free' political debate and criticism on al-Jazeera. *The Opposite Direction*, one of the most popular talk shows on the channel, was a first in the Arab world for bringing guests who would be encouraged not only to disagree with each other, but to do it seemingly without restraint.[18]

Scholarly debate on satellite television in the Arab world has since become concerned with television's potential as a democratizing tool. Al-Hail[19] and Amin[20] have written that television strengthened civil society in the Arab world, while Marc Lynch argues that al-Jazeera is opening up a space for competing voices that encourages questioning of the status quo in the region.[21] He also shares Jon Alterman's stance that Arab satellite television has created a sense of a shared Arab destiny[22], saying that satellite television 'has dramatically affected conceptions of Arab and Muslim identity, linking together geographically distant issues and placing them within a common Arab "story"'.[23] In this sense, the debate follows the classical media development approach that was prevalent in the 1980s and early 1990s. Vicky Randall's discussion back in 1993 of television in the Third World illustrates this point; she argued:

> In so far as Third World leaders have attempted to create and impose their own "political imaginary" upon their people, through a monopolistic control of the mass media, international media by providing alternative and conflicting sources of information have steadily sabotaged such a project.[24]

---

17   Ibid.

18   F. Al-Kasim, '*The Opposite Direction*: A Program which Changed the Face of Arab Television', in Zayani, *The Al Jazeera Phenomenon* (note 1) 93–105.

19   A. Al-Hail, 'The Age of New Media: The Role of Al-Jazeera Satellite TV in Developing Aspects of Civil Society in Qatar', *Transnational Broadcasting Studies* 4/ Spring (2000) at <http://www.tbsjournal.com/Archives/Spring00/Articles4/Ali/Al-Hail/al-hail.html>.

20   H. Amin, 'Satellite Broadcasting and Civil Society in the Middle East: The Role of Nilesat', *Transnational Broadcasting Studies*, 4/Spring (2000) at <http://www.tbsjournal. com/Archives/Spring00/Articles4/Ali/Amin/amin.html>.

21   M. Lynch, *Voices of the New Arab Public: Iraq, Al-Jazeera, and Middle East Politics Today* (New York: Columbia University Press 2006).

22   J. Alterman, 'Transnational Media and Social Change in the Arab World', *Transnational Broadcasting Studies* 2/Spring (1999) at <http://www.tbsjournal.com/ Archives/Spring99/Articles/Alterman/alterman.html>.

23   Lynch (note 21) p. 4.

24   V. Randall, 'The Media and Democratisation in the Third World', *Third World Quarterly* 14/3 (1993) 625–46, p. 635.

She goes on to praise the 'international media's' positive role on democracy in the region.

But al-Jazeera's potential to play an active role in espousing political change in the region was limited. Not only was al-Jazeera itself banned from reporting from a number of Arab countries whose governments were less than happy with its criticism of their regimes, but al-Jazeera also operated within political structures that suffocated any potential of the media to translate rhetoric into action. The transformation of Arab television from lapdog to watchdog was only superficial. As Rami Khouri commented back in 2001,

> Media activities in our region are still totally divorced from the political processes. An Arab viewer who might change his or her mind because of something they saw on television has no effective means of translating their views into political action or impact. For the political decision-making systems in most Arab countries are preconfigured to maintain a pro-government, centrist majority that allows more and more debate and discussion of important issues, but maintains real decision-making in the hands of small elite groups.[25]

Al-Jazeera's coverage of the intifada for example, graphic and supportive as it was, did not force any Arab government to change its stance towards the Palestinian problem. Nor did its criticism of various Arab regimes result in the stepping down of any rulers. If anything, the channel faced accusations of sensationalism and voyeurism, especially in its decision to air graphic images of the dead and wounded. However, al-Jazeera was to receive direct political accusations after September 11, when it became the medium of choice for Osama bin Laden whenever he wanted the world to see and listen to his pre-recorded video messages. Those charges came from the American administration that accused al-Jazeera of acting as a mouthpiece for and supporting al-Qaeda. The accusations led to the arrest of one of al-Jazeera's journalists, Taysir Alluni, on terrorism charges in Spain, the closing down of al-Jazeera's office in Iraq after the American invasion of the country and the refusal by a number of American officials to grant interviews to al-Jazeera.[26] Al-Jazeera had been the only channel allowed into Afghanistan under the rule of the Taliban, opening its office in Kabul in 2001.[27] As *The Economist* wrote, '[b]eing Arab and Muslim, its reporters gained privileged access to the losing side on the Afghan front', allowing the channel to broadcast a different perspective on the war on terror from the American channels.[28] Al-Jazeera's office

25 Quoted in J. Campagna, 'Arab TV's Mixed Signals', *Foreign Policy* 127/ November-December (2001) 88–9, p. 89.

26 H. Miles, *Al-Jazeera: How Arab TV News Challenged the World* (London: Abacus 2005).

27 S. Tatham, *Losing Arab Hearts and Minds: The Coalition, Al-Jazeera and Muslim Public Opinion* (London: Hurst and Company 2006).

28 *The Economist* (note 16) p. 24.

was hit by an American bomb on 12 November 2001, with many of its workers believing that the hit was a deliberate attempt by the American administration at silencing the channel.[29] A subsequent hit on its office in Baghdad in 2003 only worked to emphasize this belief. The war on terror, then, consolidated al-Jazeera's position as a political actor in the Middle East with a role that extends beyond the immediate intra-Arab political sphere.

Satellite Television as a Political Battlefield

The war on terror also challenged the primacy of al-Jazeera in Arab satellite television 24-hour news. The invasion of Iraq proved to be the greatest catalyst. News of a looming war on Iraq led Saudi-owned MBC to launch its planned news channel early. Al-Arabiya began broadcasting on 3 March 2003. From the start, MBC has marketed al-Arabiya as an alternative to al-Jazeera. Before the channel's launch, MBC had announced that al-Arabiya 'will have a non-sensationalist approach and should be perceived by the Western world as more balanced than Al Jazeera'.[30] This stance continued after the war. Al-Arabiya's director of operations Sam Barnett said in March 2004: 'There was a perception that Arab media was dominated by Al-Jazeera and that they had a certain line that was populist, heading towards sensationalist, and that there was a gap for a more considered and less sensationalist approach'.[31] Al-Arabiya maintains this stance today. The channel celebrated its fifth anniversary in 2008. To commemorate the occasion, the channel launched a series of adverts titled 'Al-Arabiya shook the world', with its media relations manager Nasser Al-Sarami asserting al-Arabiya's 'loyalty to its neutral journalistic stance that does not feed on viewers' instincts and emotions'.[32]

Marc Lynch argues that al-Arabiya initially imitated al-Jazeera in its coverage of the Iraq war in order to gain audiences.[33] However, in contrast to al-Jazeera's clear anti-war stance, al-Arabiya chose to be more ambivalent during the early days of the war. Steve Tatham compares the coverage of the fall of the Saddam statue on 10 April 2003 on the two channels and shows a clear difference in stance towards this event. While al-Jazeera covered it with a degree of lament, al-Arabiya's coverage was more hesitant. For example, al-Jazeera questioned whether the event was one of a 'foreign invader chopping off another head? Does

---

29   E. Bessaiso, 'Al Jazeera and the War in Afghanistan: A Delivery System or a Mouthpiece?', in Zayani, *The Al Jazeera Phenomenon* (note 1) 153–170.

30   S. Postlewaite, 'Al Jazeera Rival Hits Middle East', *Advertising Age* 74/5 (3 February 2003) p. 1.

31   Quoted in Tatham (note 27) p. 74.

32   Quoted in H. Nayouf, 'Al-Arabiya Begins its Sixth Year With a New Look and "News as Big as the Event"', *Al-Arabiya website* (19 February 2008) at <http://www.alarabiya.net/articles/2008/02/19/45858.html>.

33   Lynch (note 21).

the world usually use this method to honour national martyrs?'[34] Al-Arabiya on the other hand commented: 'now we will know if the US was really after freeing the Iraqi people, or after Iraqi territory'.[35]

After the war, al-Arabiya changed its coverage into a more pro-American one 'in order appeal both to the United States and to Arab elites threatened by al-Jazeera's powerful critiques'.[36] The appointment of Abdul Rahman al-Rashed as Chief Editor in 2004 is often cited as the reason behind al-Arabiya's change of stance; both *The Economist*[37] and Hugh Miles[38] for example explicitly refer to him as being 'pro-American'. However it can be argued that this change in direction could be a response to heavy criticism by the American government of al-Arabiya's early coverage of the Iraq war; Marc Lynch explains:

> In July 2003, Deputy Defense Secretary Paul Wolfowitz accused al-Jazeera and al-Arabiya of incitement to violence against coalition forces. In September 2003, Mustafa Barzani (then holding the rotating presidency of the IGC [Interim Governing Council]) ordered the closure of al-Jazeera and al-Arabiya, and in December expelled al-Arabiya for two months for playing an audiotape from Saddam Hussein. In November, after the IGC raided al-Arabiya's offices and banned its broadcasts, Rumsfeld described al-Jazeera and al-Arabiya as 'violently anti-coalition' and claimed to have seen evidence that the Arab stations were cooperating with insurgents.[39]

This change in stance was visible on the screen. *The Economist* cites the coverage of the November 2004 offensive on Fallujah by US marines as an example of al-Arabiya's divergent approach, compared with al-Jazeera:

> While al-Jazeera focused on civilian deaths and heroic resistance, al-Arabiya pictured the storming of a terrorist haven. Before Iraq's election, the Dubai channel broadcast saturation get-out-the-vote advertising, as well as a four-part exclusive interview with the interim prime minister, Iyad Allawi.[40]

This divergence led to a war of words between the two channels. Al Rashed defended al-Arabiya by saying 'We attract liberal-minded people … Jazeera attracts fanatics', while al-Jazeera's news editor Ahmed al-Sheikh responded

---

34 Al-Jazeera broadcast, quoted in Tatham (note 27) p. 138.
35 Al-Arabiya broadcast, quoted in ibid., p. 139.
36 Lynch (note 21) p. 23.
37 *The Economist* (note 16).
38 Miles, *Al-Jazeera.*
39 Lynch (note 21) 213–4.
40 *The Economist* (note 16) p. 25.

by saying that al-Arabiya is 'losing legitimacy fast ... We've got to uphold our principles'.[41]

But the battle between the two channels is not only driven by their competition in the media field. It has also been argued that both channels are affected by the respective governments of their financiers. While neither channel is a state-owned one, the reliance of al-Jazeera on the Emir of Qatar for funding and al-Arabiya's being part of a network belonging to a relative of the Saudi royal family have had an impact on their relationship as well as on their content. In recent years, Qatar has started playing a more prominent role in Arab politics. As Zayani notes, 'Qatar has exercised active diplomacy primarily by playing a mediating role in regional disputes'.[42] The state has for example participated in attempts at resolving the crisis in Sudan, and has offered to be a mediator in the conflict between Palestine and Israel. Zayani argues that al-Jazeera 'fits in with Qatar's attempt to play an active role in regional politics and to achieve regional influence'.[43] El-Oifi moreover argues that al-Jazeera's pan-Arab identity serves the political aims of Qatar to forge a sense of pan-Arab belonging that nevertheless emphasizes a Qatari national one;[44] in doing so, Qatar has entered a political rivalry with Saudi Arabia, traditionally the leading Arab state in the Gulf that has the most influence over pan-Arab politics. This rivalry on the ground parallels that between the channels backed by the two countries.[45] The channels' own coverage of Qatar and Saudi Arabia does little to dispel this theory. Al-Jazeera rarely criticizes the state of Qatar, while al-Arabiya is careful in its coverage of Saudi Arabia. In this sense, both channels seem to follow what Kraidy and Khalil call the 'anywhere but here' stance, whereby 'each channel takes the liberty to criticize all countries and policies except the country in which that channel is based or which finances its operations, and to focus on transnational issues to the detriment of local and national issues'.[46]

Satellite Television and Public Diplomacy

Despite their rivalry, al-Jazeera and al-Arabiya played an important role in the coverage of the Iraq war. The presence of those two channels during the war meant that the United States could no longer control the flow of images and information

---

41   Ibid.

42   M. Zayani, 'Introduction – Al Jazeera and the Vicissitudes of the New Arab Mediascape', in Zayani, *The Al Jazeera Phenomenon* (note 1) 1–46, p. 13.

43   Ibid., p. 12.

44   M. El Oifi, 'Influence Without Power: Al Jazeera and the Arab Public Sphere', in Zayani, *The Al Jazeera Phenomenon* (note 1) 65–79.

45   M. Fandy, *(Un)Civil War of Words: Media and Politics in the Arab World* (Westport: Praeger Security International 2007).

46   Kraidy and Khalil, 'The Middle East' (note 6) p. 81.

from Iraq, as – despite their differences – the channels' representations of the war highlighted different angles from those of the American media.[47] In particular, both channels highlighted the human dimension of the war that was often ignored in the Western media. The American government's response to this was the launch of its own Pentagon-supported Arabic news satellite channel, al-Hurra. Al-Hurra was created with the purpose of presenting the American government's point of view directly to the Arab audience, especially the Iraqi audience. It was established on the premise that al-Jazeera's news in particular is too sensational and biased against the United States.

Penny Von Eschen writes that al-Hurra is an example of American 'perception management' stemming from the United States' take on people in the Middle East as duped and lacking political agency.[48] It is on this premise, and on the assumption of the universality of American modes of political communication, that the United States government conceived al-Hurra as a tool of political change in the Arab world.[49] The station's name means 'the free one', and aims to send a message of fairness and objectivity to its intended audience, while also referencing the promise of 'liberation' that the invasion of Iraq was supposed to bring about. Even the channel's ident has been chosen to convey this sense of 'liberty'. The ident, which runs on the screen periodically, shows a vast white landscape dominated by the image of multicolored Arabian horses running freely, again connoting the sense of freedom that the United States is supposed to bring the Arab world through its foreign policies. Al-Hurra also presents the American point of view in its coverage of events in the Middle East. The iconic moment of the destruction of Saddam's statue in Baghdad in 2003 was framed by al-Hurra as one of liberation, in sharp contrast to the stance taken by al-Jazeera, where it framed the event as within a war *on* Iraq.[50] However al-Hurra has failed in its mission and is one of the least watched and trusted satellite stations in the Arab world.[51] Al-Hurra has failed in challenging the position of al-Jazeera, or in convincing Arab audiences that it is a credible source of information. Consequently, the United States government is now seeking less overt methods of public diplomacy in the Middle East.

But al-Hurra set a precedent. The use of satellite television to communicate to and from the Arab world has since evolved with the recognition of the need

47   Lynch (note 21).

48   P. Von Eschen, 'Enduring Public Diplomacy', *American Quarterly* 57/2 (2005) 335–43.

49   N. Patiz, 'Radio Sawa and Alhurra TV: Opening Channels of Mass Communication in the Middle East', in W. Rugh (ed.), *Engaging the Arab and Islamic Worlds through Public Diplomacy: A Report and Action Recommendations* (Washington DC: Public Diplomacy Council 2004) 69–89.

50   M. Zayani and M. Ayish, 'Arab Satellite Television and Crisis Reporting: Covering the Fall of Baghdad', *International Communication Gazette* 68/5–6 (2006) 473–79.

51   F. Braizat, *Revisiting the Arab Street* (Amman: Center for Strategic Studies, University of Jordan 2005) at <http://www.jcss.org/SubDefault.aspx?PageId=55&PubTyp e=1>.

to reach out to audiences beyond one's own. In 2004, Iran launched al-Alam, an Arabic-language news channel aimed at its neighboring audiences. In 2007, Russia launched another Arabic-language station, Rusiya al-Yaum, to present its own point of view to those same audiences.[52] The BBC has also created an Arabic-language television station which is backed by the UK's Foreign and Commonwealth Office. The FCO's involvement in the affairs of the BBC is often downplayed, but the creation of the Arabic language channel is part of the UK government's public diplomacy efforts in the Arab world. In contrast, in 2007 al-Jazeera launched an English-language channel, al-Jazeera International. Al-Jazeera International's content is differently selected, framed, and presented from that of its Arabic counterpart as it is geared towards Western and English-speaking Asian audiences.

The presence of those channels is indicative of the international nature of geopolitics in the Middle East. Key political players from within and outside the region are vying for space in the Arab televisual landscape. At the same time, the launch of al-Jazeera International is an example of this process in reverse. As argued earlier, the state of Qatar is playing an increasing role in politics in the Middle East, but it also presents itself to the West as a political mediator. Its membership of the Security Council, donations to rebuild villages in the Lebanese South after the 2006 war, and involvement in local Palestinian politics attest to its ambitions in the international political arena, since all those conflicts are not limited to their geographical locality. If we are to accept Qatar's influence on al-Jazeera, it would not be surprising for Qatar to seek to indirectly present a favorable image of itself through al-Jazeera International. This is not done through representing Qatar on the screen; it is the mere positive association between Qatar and a channel that has recruited high-profile journalists (including Western ones), relies on glossy images and uses moderate language in its reporting that can be seen to have a favorable impact on Qatar's reputation.

The accelerated launch of Al-Arabiya and the mushrooming of television stations aimed at the Arab world is indicative of the primacy of the image in the age of the war on terror. September 11 was a highly visual event. It marked a change in the global television landscape, and even in geopolitical war tactics. It confirmed the power of the image, and the impact of staging events for the camera. Although the invasion of Iraq carried less iconic images than those of September 11, it is still remembered in visual terms. The destruction of the statue of Saddam in Baghdad; the capture of Saddam Hussein in a hole; and the Abu Ghraib photographs are memorable visual moments in the war. The Iraq war has proven that contemporary warfare is incomplete with a comprehensive information management strategy that takes into consideration the role of images in general, and satellite television in specific. Geopolitics today is *seen*.

---

52   Reuters, 'Russia Launches Arabic TV News Channel', *Khaleej Times Online* (4 May 2007) at <http://www.khaleejtimes.com/DisplayArticleNew.asp?xfile=data/ theworld/2007/May/theworld_May117.xml&section=theworld>.

The Internationalization of Local Conflict

Satellite television in the region both reflects and reinforces the multi-scalar nature of political conflict in the Arab World. Several satellite stations are simply duplicates of official terreresterial broadcasting organizations (e.g. Egypt, Sudan and several other countries fall into this category). As such they convey the official ideology of the state to an international audiences as well the domestic.

Lebanon presents perhaps the clearest example of the role of satellite television as a participant in political conflict. Following the assassination of former Prime Minister Rafic Hariri in 2005, Lebanese satellite television stations were divided into competing anti-Syrian (Future TV, LBC) and pro-Syrian (Al-Manar, NBN, New TV) camps. The stations in the first camp are owned by the family of Rafic Hariri himself in the case of Future TV, and, in the case of LBC, partly by those affiliated with the Maronite Kataeb and Lebanese Forces parties – all of which have formed the pro-government '14 March coalition'. The stations in the second camp are owned by Hizbullah (Al-Manar), Hizbullah's ally Shiite group Amal (NBN), and a rival of Hariri's (New TV). Along with the Free Patriotic Movement political party led by General Michel Aoun, the latter groups formed the anti-government '8 March coalition'. Thus, the competing television stations represented the agendas of the political parties clashing on the ground.[53]

In Lebanon the television landscape is therefore deeply fragmented between the competing factions. Satellite television provides a means by which they can project their world views and corresponding agendas to wider audiences. Both camps have been engaging in a televised battle for legitimacy, using political events to appeal to the Lebanese people, while also defending their positions vis-à-vis the wider Arab audience. For example, the camps have engaged in a battle over who is the 'true' representative of 'all' the Lebanese people. The stations in the anti-Syrian camp have constructed the Hariri assassination as an event concerning all Lebanese. Furthermore it protrayed the attribution by the UN of Syrian involvement as a victory for the Lebanese people and their sovereignty. On the other hand, NBN and Al-Manar have chosen to focus on another event. Hizbullah's self-declared 'victory' in the July 2006 war is constructed by these channels as being for all Lebanese, not just Lebanon's Shiites. These events were still being used with equivalent force three years after Hariri's assassination, and two years since the July war. The events become examples of how '[t]he past plays an authenticating and legitimizing role' in the struggle to represent the Lebanon.[54] This struggle is being played out not just in the Lebanon but across the Arab World and beyond.

---

53   L. Khatib, 'Television and Public Action in the Beirut Spring', in N. Sakr (ed.), *Arab Media and Political Renewal: Community, Legitimacy and Public Life* (London: I.B. Tauris 2007) 28–43.

54   Y. Suleiman, *The Arabic Language and National Identity* (Edinburgh: Edinburgh University Press 2007) p. 38.

Both camps relied on different symbols and visual codes in their appeal to audiences. Al-Manar alternated between regular images of devastation from the 2006 war and pride in Hizbullah's 'victory', with the images becoming a visual signifier of the need for the maintenance of Hizbullah as a military organization that serves a 'defensive' role in Lebanon. Following Hariri's assassination, Future Television carried an on-screen counter tallying the number of days since the assassination. The on-screen counter served as a constant reminder of his mysterious death. Michel Foucault argued that memory is 'a very important factor in struggle ... if one controls people's memory, one controls their dynamism'.[55] In Lebanon, both camps are aware of television's potential for establishing hegemonic understandings of past events.

In 2007, another Lebanese political group joined in the satellite television battle. The Free Patriotic Movement launched Orange Television (OTV) in autumn 2007. The Free Patriotic Movement's alliance with Hizbullah revolves around lobbying for Aoun to become president of Lebanon as well as maintaining Hizbullah as a paramilitary group. OTV started airing in the period leading up to the Lebanese presidential elections that were supposed to take place in November 2007. The use of visual codes to transmit political messages is at the heart of OTV, with the station's name being that of the signature color of the Free Patriotic Movement. The urgency with which OTV was created stems from the fragmentation of the media space in Lebanon, where if a political party does not own a television station, its views are marginalized. This is because there is no space in the Arab television landscape where competing views are given an equal share.

Conclusion

The television landscape of the Arab World has become far more diverse with the advent of satellite broadcasting. Moreover, this poses a challenge to the dominance of Anglo-American media channels. All Arab satellite channels have been involved in covering the Iraq war, and despite their differences, they have all presented a degree of criticism of American actions on Arab soil that far exceeds that on American television. Adel Iskandar for example notes how Fox News Channel's coverage of the invasion of Iraq contained 'little to no footage of civilian casualties ... and infrastructural damage in Iraq was shown primarily via long-distance footage ... with voiceover military reports proclaiming accuracy in striking strategic targets'.[56] It was the Arab television stations that first showed

---

55   M. Foucault, 'Interview: Film and Popular Memory', *Radical Philosophy* 11 (1975) 24–29, p. 28.

56   A. Iskandar, '"The Great American Bubble": Fox News Channel, the "Mirage" of Objectivity, and the Isolation of American Public Opinion', in L. Artz and Y. Kamalipour (eds), *Bring 'Em On: Media and Politics in the Iraq War* (Oxford: Rowan & Littlefield 2005) 155–73, p. 162.

the impact of the war on the Iraqi people. A similar situation occurred with the July 2006 war between Israel and Hizbullah, where the scale of the human and infrastructural tragedies in Lebanon caused by the war found a voice primarily in the Arab media.

However, the existence of any kind of real political dialogue through Arab television stations remains unattained. Al-Jazeera and al-Arabiya often present different versions of the same events that are indicative of their clashing political stances. For example, during the July 2006 war, al-Arabiya identified Hizbullah as a 'Lebanese Shiite' group, whereas al-Jazeera simply referred to it as a 'Lebanese' group, thereby creating clear, clashing frameworks of (il)legitimacy surrounding Hizbullah's actions and motives in the war. Competing stations closely monitor and respond to each other, but do so to discredit the other, rather than engage with them. An example is al-Manar's coverage of the street riots that took place in Lebanon in January 2007, where the station often began its broadcasts by quoting the news reports of Future Television and then branding them lies. News coverage thus has become an exercise in political strategy.

In this febrile context, it is not surprising that outside political actors have jumped on the bandwagon of using television as a mouthpiece to address the Arab world. The presence of al-Jazeera International as well as Iranian, British, Russian and American satellite television stations broadcasting in Arabic has complicated what is meant by 'Arab' satellite television, and confirmed television's role as a participant in political conflict. But even in the case of satellite television stations owned by Arabs that broadcast in Arabic, the situation is complex. The Arab satellite television landscape is one of contention, indicative of power struggles within the Arab world and between it and outside forces. Satellite television stations engaged in news reporting act as mouthpieces for clashing political actors whose primary motive is the propagation of messages favourable to themselves. In this situation, a genuine political dialogue through television will remain difficult.

Chapter 12

Maranatha! Premillennial Dispensationalism and the Counter-Intuitive Geopolitics of (In)Security

Jason Dittmer

Cheer up Saints, it's gonna get worse![1]

Introduction

Evangelical Christians as a group have been perceived to be of increasing geopolitical importance over the past several decades, but most especially during the presidential administration of George W. Bush. During the Bush presidency (2001–2009) a number of theories have been advanced regarding the linkage between this constituency and many of Bush's more controversial policies in the Middle East; including support for Israel and regime building in Iraq and beyond. As Esther Kaplan had noted, 40 per cent of Bush's electoral support was composed of Christian evangelicals who enjoyed close links with congressional Republican leaders such as Tom de Lay.[2] This chapter is less concerned with these theories than how geopolitical events interact with evangelical theology to produce feelings of security and insecurity among this core constituency. It will also examine the role that the evangelical Christian Internet community plays in disciplining participants' subjectivity and bolstering their emotional security. Further, this case study examines the role of religion in providing emotional security in times of perceived danger by documenting how a subset of evangelical Christians who believe in the eschatology called premillennial dispensationalism gain a sense of security (or fail to do so) by interpreting geopolitical events in a communal setting.

Premillennial dispensationalists are relevant theoretically because of their belief in a particular reading of the Bible, in which global geopolitical affairs will generally deteriorate until Jesus Christ returns to fulfill the prophecies and establish a perfectly just kingdom that will last for a millennium. Thus, their geopolitical perspective is pessimistic regarding the possibility of peace, environmental sustainability, and other subjects of the international agenda. Paradoxically, this means that geopolitical 'bad news' is really affirmation of these believers' prophetic

---

1   Rapture Ready discussion board post 449081 (16 February 2008). Posts to this board are hereafter referred to by their number and date.

2   E. Kaplan, *With God on Their Side* (New York: New Press 2005) p. 25.

beliefs, and thus a source of emotional security. The title of this chapter references 'Maranatha', an Aramaic word that roughly translates as 'Come, O Lord'. Some premillennial dispensationalists use this word as an individual performance of this communal security to be found in the end of the corrupt world and Jesus's subsequent return.

The first section of this chapter outlines the basic beliefs of premillennial dispensationalism, showing how many adherents to this eschatology invert traditional notions of security. Subsequently, this chapter illustrates the ways in which the *Rapture Ready* online discussion board (one of the oldest and largest evangelical websites on the Internet) can be understood to discipline its participants and constitute a coherent theological community. The penultimate section of the chapter will report and contextualize the results of ethnography conducted via internet discussion threads continually updated on the *Rapture Ready* website, in which participants discussed their sense of security in relation to current events in early 2008. The chapter concludes by considering the implications of the analysis and potential avenues for future research.

Premillennial Dispensationalism and the Rapture Ready Message Board

Premillennial dispensationalism refers to a set of evangelical Christian beliefs about the structure of time and the end of the world. This particular interpretation (one of many about the 'End Times') relies on a 'literal' reading of the prophetic books of the Bible (most commonly Daniel, Ezekiel, and Revelation), in which a series of relationships (known as dispensations) between humanity and God are entered into, with humanity failing to uphold its end of the dispensation each time. The dispensations are: 'Innocency (before the Fall), Conscience (Fall to the Flood), Human Government, Promise (Abraham to Moses), Law (Moses to Christ), Grace (the church age), and Kingdom (the millennium).'[3] According to dispensationalist thought, we are currently in the church age, waiting for the beginning of the millennial kingdom, and as we shall see, it is the moment of this shift from one dispensation to another that preoccupies the followers of this belief system.

According to this theology, the end of the previous era, the dispensation of Law, came when the Jews did not recognize Jesus as their messiah. With this breaking of the dispensational covenant, God set the Jews aside for the time being in order to work with the Gentile church. However, the promise of God to his Chosen People will not be broken, and God will return to them in the final dispensation (the millennial kingdom) to give them spiritual guidance as well as actual political guidance. As Evangelical historian Timothy Weber puts it:

> Until Messiah's [second] coming, however, God's earthly people must suffer Gentile domination, prophesied by Daniel. This Gentile hegemony would end at

---

3   T. Weber, 'The Dispensationalist Era', *Christian History* 18/1 (1999) 34–7, p. 35.

the coming of Messiah, 70 weeks after one of the Gentile rulers issues a decree allowing the Jews to return to Jerusalem to repair its broken walls. But when the Jews rejected Jesus as their Messiah, God suspended the prophetic timetable at the end of Daniel's sixty-ninth week and began building a new and heavenly people: the church.[4]

As the above quote shows, this teleological understanding of history is tied to specific geopolitical events, such as the return of the Jews to Jerusalem (i.e., the Balfour Declaration and the independence of Israel in 1948). The important point for this chapter regarding premillennial geopolitics is that many premillennial dispensationalists are looking for 'signs of the times' (such as earthquakes, terrorism, school shootings, etc.) that indicate that God is getting ready to re-start the prophetic timetable, finish Daniel's seventy weeks of Gentile domination, end the Church Age, and begin Jesus's one thousand-year reign over the earth (the millennial Kingdom). The end of the Church Age will be marked by seven years of natural disaster and warfare, usually called the Tribulation. The Tribulation will devastate the population of the Earth, and serves as a final warning to repent and become born again in Jesus before he returns at the head of a glorious army to sit in judgment.

I have dealt with the contemporary geopolitics of premillennial dispensationalism in previous publications and thus will not cover that ground again here.[5] However, simply put, premillennial dispensationalism is the most influential form of American religious apocalyptic thought. Originally imported from the United Kingdom in the nineteenth century, premillennial dispensationalism has historically thrived in the American Northeast and Midwest, co-existing with (and in some ways in resulting from) Northern liberal Protestantism. Over time however, premillennial dispensationalism has found some common cause with Southern conservatives via informal collaboration and has thus risen in influence. Although adherence to this particular theology (or any other) is hard to measure, estimates are that between 30–44 per cent of Americans believe in the Rapture, of which the strong majority are premillennial dispensationalists.[6]

The Tribulation is important for premillennial dispensationalists because they believe it will be preceded by the Rapture. The Rapture is a highly contested notion, as many Protestants disagree with the premillennial dispensationalists, believing instead that the Rapture will occur at the end of the millennial Kingdom,

---

4   Ibid.

5   J. Dittmer, 'Of Gog and Magog: The Geopolitical Visions of Jack Chick and Premillennial Dispensationalism', *ACME: An International E-Journal for Critical Geographies* 6/2 (2004) 278–303; J. Dittmer, 'The Geographical Pivot of (the End of) History: Evangelical Geopolitical Imaginations and Audience Interpretation of Left Behind', *Political Geography* 27/3 (2008) 280–300.

6   D. Wojcik, *The End of the World as We Know It: Faith, Fatalism, and Apocalypse in America* (New York: NYU Press 1999).

in the middle of the Tribulation, or that there is no Rapture. As this chapter is focused on *premillennial* beliefs, it is worth quoting the premillennial definition of the Rapture as given on the *Rapture Ready* website:

> The rapture is an event that will take place sometime in the near future. Jesus will come in the air, catch up the Church from the earth, and then return to heaven with the Church. The Apostle Paul gave a clear description of the rapture event in his letters to the Thessalonians and Corinthians.

> "For the Lord himself shall descend from heaven with a shout, with the voice of the archangel, and with the trump of God: and the dead in Christ shall rise first: Then we which are alive and remain shall be caught up together with them in the clouds, to meet the Lord in the air: and so shall we ever be with the Lord. Wherefore comfort one another with these words." (1 Thess, 4:16–18)[7]

Thus, 'real' Christians will be able to avoid the Tribulation by joining Jesus in heaven, avoiding the calamities and strife that will befall those left behind. This is the source of security for believers – it allows them to view the end of the world with a sense of critical distance and personal protection.

This chapter is based on ethnographic research conducted on those who posted on the *Rapture Ready* message board (http://www.raptureready.com) over a one-month period from 25 January–25 February 2008.[8] The discussion threads came from the 'Prophecy and The End Times' section of the discussion boards. One hundred fifty-six threads were coded, with a focus on expressions of security or insecurity in response to current events. This research differs from other ethnography in that it was not carried out in a face-to-face material setting, and in some ways it has more similarities to discourse analysis than ethnography. However, given the growth of online communities as significant players in the religio-political realm this methodological hybridity can be seen to reflect these encounters' hybrid constitution as both community and text.[9] Dias's work on pro-anorexia web fora in some ways parallels this research.[10] Just as the pro-anorexia individuals were atomized in geographic space and were only able to find community in cyberspace, the eschatologically-minded premillennial dispensationalists are themselves a subset of a subset of a subset. Often they abandon church-going because they find all modern churches apostate from 'real' Christianity. Thus, the virtual church

---

7   <http://www.raptureready.com/rr-pretribulation-rapture.html>.

8   This is a method pioneered in geography by K. Dodds, 'Popular Geopolitics and Audience Dispositions: James Bond and the Internet Movie Database (IMDb)', *Transactions of the Institute of British Geographers* 31/2 (2006) 116–130.

9   See for instance moveon.org, or the virtual *umma* described in Retort, *Afflicted Powers: Capital and Spectacle in a New Age of War* (London and New York: Verso 2005).

10   K. Dias, 'The Ana Sanctuary: Women's Pro-Anorexia Narratives in Cyberspace', *Journal of International Women's Studies* 4/2 (2003) 31–45.

represented by *Rapture Ready* is, for many, their primary connection to a religious community.

Fear, Security, and the Rapture Ready Message Board

One purpose of this chapter is to turn the discussion of post-9/11 geopolitics from the field's traditional focus on the actions of elites to the more banal experiences of everyday life as lived by everyday people, a move advocated by both scholars of popular and feminist geopolitics.[11] One of the most important ways in which this move can contribute to the literature is by breaking down the false monolith of 'national interest' and showing how, for example, increased securitization of public spaces has differential outcomes among social groups (with the burden in the US and UK perhaps falling hardest on young male Muslims).[12] Most accounts of empire function with a normative empire/colonized dichotomy, with the imperial state(s) inherently advantaged as a result.[13] There have been few studies discussing the negative impacts of empire on imperial citizens, such as chronic fear.[14] However, this is beginning to change, as are the scales at which fear is analysed:

> [T]here is an uneasy yet taken-for-granted assumption that fear-provoking incidents take place, and fear-inducing discourses circulate, at one (global) scale/space, inducing people to become fearful at other (more local) sites. This received wisdom is, however, at odds with the recent 'emotional turn' in social and economic research, which recognises the complexity, situatedness, sociality, embodiment and – critically – constitutive qualities of emotional life. Fear does not pop out of the heavens and hover in the ether before blanketing itself across huge segments of cities and societies; it has to be lived and made.[15]

---

11   For example, J. Hyndman, 'Towards a Feminist Geopolitics', *The Canadian Geographer* 45/2 (2001) 210–22; J. Hyndman, 'Mind the Gap: Bridging Feminist and Political Geography Through Geopolitics', *Political Geography* 23/3 (2004) 307–322; F. MacDonald, 'Geopolitics and "the Vision Thing": Regarding Britain and America's First Nuclear Missile', *Transactions of the Institute of British Geographers* 31/1 (2006) 53–71; R. Pain and S. Smith (eds), *Fear: Critical Geopolitics and Everyday Life* (Aldershot: Ashgate 2008); Dodds (note 7).

12   K. Hörschelmann, 'Youth and the Geopolitics of Risk after 11 September 2001', in Pain and Smith (note 10) 139–152.

13   For example, M. Mann, *Incoherent Empire* (London: Verso 2004).

14   But see C. Johnson, *Blowback: The Costs and Consequences of American Empire* (New York: Henry Holt 2004).

15   R. Pain and S. Smith, 'Fear: Critical Geopolitics and Everyday Life', in Pain and Smith (note 10) 1–19.

An increasing focus on emotional geographies has highlighted the ways in which understandings of the relationship between environment and people impact subjects in ways that had previously been left undertheorized.[16] 'Emotions are widely understood to be contained by the psycho-social and material boundaries through which embodied persons are differentiated from one another and from their surrounding environments. Although taken-for-granted in the course of much everyday life, such boundaries are never impermeable or entirely secure.'[17] Both positive and negative emotions are associated with the transgression of these psycho-social and material boundaries. In the Western tradition, the management of this emotional traffic is purportedly to be regulated by the subject as an indicator of the subject's sovereignty.

Katz argues that fear and anxiety is the result of the privatization of social (in)security in the neoliberal state: 'In the thrall of insecurity, individuals as much as social formations are all the more vulnerable to fear-mongering, the machinations of banal terrorism and 'terror talk'. They tend to turn inward in response.'[18] Maintaining a sense of calm in a hyper-mediated world of conflict (the symbol of individual sovereignty) can be difficult. This emotional imperative was seized upon by the George W. Bush administration, which rather than declare a 'Global War on Terrorism' instead declared a 'Global War on Terror' in recognition of the political power to be wielded by those who can alleviate perceptions of insecurity.[19]

Disciplining the Premillennial Subject

It is worth considering here Foucault's ideas of bio-power, through which he argued that subjects are constituted through self-discipline and the discursive construction of disciplinary norms, such as rationality, heterosexuality, etc.[20] Thus, having 'correct' (or governed) emotions is seen as constitutive of the 'healthy' subject, with the inability to govern one's borders pathologized and seen often as grounds for medico-state intervention. Scaled up, the same argument can be made for states – with the same questions of boundary maintenance, sovereignty, and security being asked of 'failed states' and other sites of intervention by an 'international community' that relies on medical metaphor to justify intervention see also Chapter

---

16   For example, J. Davidson, L. Bondi and M. Smith, *Emotional Geographies* (Aldershot: Ashgate 2005).

17   L. Bondi, J. Davidson and M. Smith, 'Introduction: Geography's "Emotional Turn"', in Davidson et al. (note 15) 1–16, p. 7.

18   C. Katz, 'Me and My Monkey: What's Hiding in the Security State', in Pain and Smith (note 10) 59–70, p. 61.

19   D. Cowen and E. Gilbert, 'Fear and the Familial in the US War on Terror', in Pain and Smith (note 10) 49–58.

20   M. Foucault, *Discipline and Punish: The Birth of the Prison* (New York: Vintage Books 1978).

2).[21] In all of these cases, it is the same ability to outwardly manifest self-discipline that demonstrates viability. The geopolitical subject (whether individual or state) is a *performance* of discursively-produced Western rationality.

By viewing premillennial dispensationalism through the disciplinary lens of Foucauldian biopolitics, it can be seen as a regime of governance, not by national elites but by religious ones intent on reining in the multitude of possible theologies and creating orthodoxy. The *Rapture Ready* discussion board can be understood as a technology allowing the social policing of premillennial dispensationalists and their beliefs. While there is no literal punishment for holding beliefs that are distinct from the board's orthodoxy, and indeed there is no requirement to participate, the discussion board members hold varying degrees of communal influence over each other, and ultimately the board administrators control the perspectives that are made visible through the website. As an example of this disciplinary action, consider the following quotes from a thread that began on the topic of Russian cruise missile tests (posters' names are pseudonyms):

> *Voice of Reason*: I'm not worried about Russia and China's recent military build-ups. Think about it, there's a new article every day about a "new missle this" or "new fighter jet that" Notice, the USA doesn't publish anything, and that's for a reason. I definitely still have confidence in my stripes and stars.[22]

> *Josiah*: That's a dangerous place to be. Our hope is in God, not the USA.[23]

In this exchange, the first poster expresses faith in a secular form of security, and is rebuked by the second poster. In actuality, these kinds of rebukes are fairly uncommon. However, this is because the performance of identity on *Rapture Ready* is often quite anxious to project confidence, with many posts beginning with a conversion narrative, or other narrative of faith, that establishes the poster's credentials as a member of the community:

> I know what you mean [about being revolted by sinful place]. I was that way when I was saved as a child, but later I got in with a bad crowd and backslid horribly for years. I repented and I am trying to lead a Christian life (not perfect though). This weekend when I was near New Orleans visiting my fiance and I went into the city while he was working to buy a gift for my roommate/bff (she loves clowns/jesters – got her a tee shirt). I used to love the French Quarter and the music and fun, but this time I did not enjoy the excursion – started feeling really repulsed and almost ill (alot of homosexuals around also) that I started practically running through the little streets to get out of there. I didn't even eat

---

21 D. Campbell, 'Geopolitics and Visuality: Sighting the Darfur Conflict', *Political Geography* 26/4 (2007) 357–82.

22 Post 417843 (26 January 2008).

23 Post 421025 (28 January 2008).

the seafood I usually have to have when I am in the area – just couldn't be down there anymore and couldn't cross the river fast enough to get away, it seemed. Just really strange for me since I used to be very worldly and "street tough", but found so much disgusted me and then I was angry at myself for going there in the first place, especially alone[.][24]

Narratives such as this illustrate the proper ways in which *Rapture Ready* participants are supposed to feel and behave. Expressions of faith serve to bolster the group identity and discipline adherents.

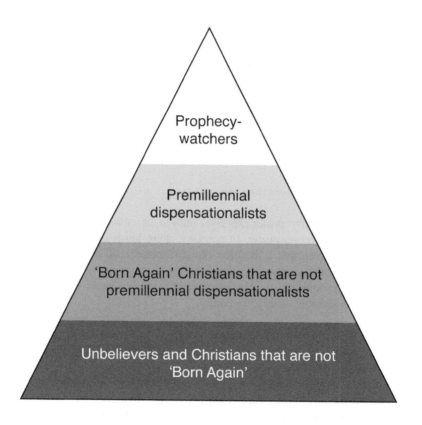

Figure 12.1 A visualization of the hierarchy of identities discursively produced on the *Rapture Ready* message board

---

24   Post 452343 (18 February 2008).

Crampton points out that categories in which individuals can be identified as threats or non-threats are necessary for the governance of populations, as well as surveillance technologies that enable that categorization to be made in a 'rational' fashion.[25] I would argue that on the *Rapture Ready* discussion board the world's population is generally subdivided into four discursive categories, each of which is labelled according to the group's theological stance in regards to Jesus Christ. These categories can be imagined as a pyramid in that they are perceived as hierarchical and the population of each group declines as you ascend the hierarchy (see Figure 12.1). However varying levels of anxiety that is sometimes non-intuitive mark each group's relationship with the posters on the discussion board.

At the base of the pyramid are unbelievers or non-'born again' Christians, those who need to accept Jesus as their savior prior to their death or the Rapture (whichever comes first). As such, they are subject to the evangelism of the premillennial dispensationalists, but are also a mild danger to premillennial dispensationalist identity because their secular (or Catholic, New Age, or Muslim, or whatever) beliefs can weaken the resolve of born again Christians. The next category of subjects are born again Christians who are not premillennial dispensationalists. The most commonly discussed groups were 'Posties ' (postmillennialists – who believe Jesus will only come at the end of the millennial kingdom, and hence there is no Rapture or Tribulation) and Preterists (those who believe that some prophecies in the Bible were fulfilled in the first century AD). These groups are technically 'saved', but do not have all the details 'correct'. Despite many beliefs held in common with the premillennial dispensationalists, they are the only group that is explicitly banned from posting on the discussion boards, as evidenced by this excerpt from the rules for posting on the website:

> *No promotion of Mid or Post Tribulation, Prewrath, Preterism, or Replacement theology.* This board is traditional Pre-Tribulation Rapture, Pre 6 Seals, Pre 70th week, and Pre-Millennial Dispensational in our position of End Times Prophecy. We also believe in a literal 7 year Tribulation period, after the instantaneous Rapture of <u>all</u> regenerated believers in unison, during which God finishes His discipline of Israel and finalizes His judgments on the unbelieving world. In order to stay within our scriptural position and keep order and peace, all discussions on these topics must follow suit. Our Lord Jesus Christ told us that we won't know the day or hour of His return and to always be ready keeping watch.[26]

While it is hard to know how many threads have been deleted for violating this prohibition given their subsequent absence from the site, there are numerous threads available in which website administrators have intervened to erase hyperlinks or other text associated with non-orthodox theologies (a note is left

---

25   J. Crampton, 'Biopolitical Justification for Geosurveillance', *Geographical Review* 97/3 (2007) 389–403.

26   <http://www.rr-bb.com/showthread.php?t=2>.

when this is done). While premillennialists hope to convert these people to their eschatology, they view this as relatively low-stakes in comparison to the conversion of unbelievers, who it is believed will go to Hell if they are not saved. Having said that, the anxiety expressed through the ardent policing of the discussion board for postmillennialism or other evangelical variants reflects a sense of their own subjecthood's vulnerability to such persuasion and thus a need to police the boundaries of acceptable speech.

Premillennial dispensationalists themselves occupy the next level of the hierarchy. This is the implied audience of the website, and as seen above, other views are screened out.[27] While many posters express hope that the website serves an evangelistic purpose, more express gratitude to the board for providing a community of like-minded Christians who feel excluded from their 'real' churches as a result of their ardent belief in the prophecies of the Bible. The very highest level of the hierarchy are the 'prophecy-watchers' – those who like to watch the news and speculate on events' eschatological meaning. This is a subset of the *Rapture Ready* community, but for obvious reasons a strong majority in the particular forum analysed in this research, 'Prophecy and The End Times'. This group did not snub non-prophecy-watchers but often expressed incredulity that others did not share their interest. Prophecy-watching was seen as not necessary for salvation but some believed they would receive extra plaudits from Jesus after their Rapture that non-prophecy-watchers would not receive. Disciplining of these prophecy-watchers was also undertaken by the website administrators, who locked any threads that speculated on the identity of the Antichrist or attempted to set dates for the return of Jesus (again, from the rules for posting):

> *No End Times Date Settings, Date Alerts, or designating specific Peace Treaties as the main event* nor quoting from others on date setting future events such as the fall of Damascus, Gog/Magog, the Rapture, the Second Coming, the Millennium, Peace Covenants, etc. or speculations on who will be the Antichrist or False Prophet. This includes linking to sites and books that also speculate. Do not set Date Alerts that even hint of date setting.[28]

Date setting and Anti-Christ spotting on the website produces anxiety because it sets up the premillennial community for mockery by outsiders.

The *Rapture Ready* discussion board thus serves as a social outlet for premillennial dispensationalists, who are often marginalized within their own churches as a result of their ardent prophecy-watching (if they even go to a church anymore – many are disenchanted because of this marginalization). However, the power inherent to the position of board administrator allows these individuals to select the discourses available on the site. As a support system for socially

---

27   For more on the implied audiences of 'New Media' see S. Livingstone, 'New Media, New Audiences?' *New Media & Society* 1/1 (1999) 59–66.

28   <http://www.rr-bb.com/showthread.php?t=2>.

atomized individuals, the website serves as a venue for the public performance of identity. It is to these public performances that this chapter now turns.

Performing (In)Security

What makes premillennial dispensationalists interesting for scholars studying notions of security is the paradoxical relationship between physical danger and their sense of security. Because their teleological perspective is of a world in inevitable decline, leading up to the much-anticipated return of Jesus, they generally view efforts to ameliorate suffering or to secure peace in conflict-ridden areas as the work of Satan. Indeed, their perception of the contemporary world as in decline serves as evidence for their faith and their teleological futurology. This link between posters' positive sense of well-being (the bolstering of their faith and the eternal security they associate with it) and the violence, disaster, and moral corruption they see around them (traditionally associated with a lack of security) can be found in their own writings, which are detailed below.

Security

Perhaps the most succinct way of describing the attitudes of the posters is by describing their use of the word 'maranatha'. As stated above, this Aramaic word translates roughly as 'Come, O Lord' and was used as a greeting among early Christians. On the discussion board, however, it is also used in a more literal sense, as a personal call to Jesus to come back soon. This desire to leave the world and be reunited with Jesus through the Rapture is something usually actively desired, as the use of 'maranatha' implies. The sense of anticipation is keen for many posters, who literally expect to be taken away at any moment:

> Here in Japan my house is surrounded by small mountains. Well this morning a little after six I was walking to the train station with snow falling when I heard a horn blow which echoed through the mountains. My heart skipped a beat or two thinking here we go.. :yeah but alas nothing.:tsk There is a buddist shrine about half a mile a way and I guess somebody decided to blow a horn for the fun of it. They have never done that before so it was a bit of a surprise. O well next time.[29]

The 'signs of the times' were studied by those contributing to the discussion board for the purpose of considering the notion of reassurance (all quotes are unmodified from their original form to maintain accuracy; please note they were typed in a casual online forum):

---

29   Post 444305 (12 February 2008).

More changes have taken place so far as stage setting for prophecy and the return of Christ since World War 2 than at any other time in history. Some of those changes are: 1. Israel is in the land in unbelief. 2. Europe is uniting. 3. The world is moving towards globalism and global government (UN, et al). 4. There is growing apostacy (of which the Emerging Church is at the top of the list). 5. Technology needed for the fulfillment of many prophecies exist now. 6. Iran is moving into its place as leader of the Moslem world. 7. Russia is rising. 8. Russia, Iran and the other countries listed in Ez. 38 are making military alliances. This has never happened before in history. I could go on and on, but those are just a few. Prophecy wasn't given to scare us to death but to motivate us as we see the day drawing near!!!![30]

These signs of the times, as stated above, did not provoke feelings of insecurity but rather the opposite:

I'm not sure people will understand this, but I find studying prophesy to be comforting. I don't know how other people can stand watching the news without knowing prophesy. When I see the horrific events happening and soon to happen I can read prophesy to remind myself that God foresaw all this and that it is under His control. It gives me such hope to think that there is going to be an end to this madness, and that we will soon be with Jesus.[31]

This sense of security is based not only on the eternal life that adherents believe they will receive following the Rapture, but also in the belief that they will escape the physical pain and deprivation associated with the Tribulation:

I think we are seeing the cooking of World War 3. I can smell it. Alot of countries are doing alot of blood shed. I believe we will see alot more before the rapture even happens. Noah was right in it, just before the ark was closed. It was wicked and I'm sure it was so bad, their family could of been killed at anytime. But once again, God protected them. God never changes. He knows exactly the moment He needs to remove His people.[32]

Thus, faith in the Rapture and in the eternal life that accompanies it left many posters emotionally secure.

However, some posters took that sense of security and connected it, despite the board administrators' admonishments against 'date-setting', with contemporary geopolitical events.[33] Sure that the end of the age is upon them, and secure in

---

30 Post 420550 (27 January 2008).
31 Post 420811 (27 January 2008).
32 Post 425120 (30 January 2008).
33 By saying that an event was of prophetic significance, posters could signal that the end is near without actually committing to an actual end date (and thus avoid having their

their position as those to be Raptured, they are excited by being able to witness prophetic events unfolding around them. This is a level of security from anxiety not talked about in other literatures, one in which the perceived 'end times' are on display as spectacle: 'So many christians act like I'm crazy because all I do is read the news and talk about it. They laugh at me and tell me not to worry about it. I can't seem to get it into their heads that I'm NOT worried! I'M EXCITED!!!!'[34]

Like this poster, others hint at the social disjunctures that often follow in the wake of their excitement:

> Today my mother told me that she is "afraid" that a certain person will be elected president, and I just told her I'm not afraid of anything. Later we were shopping and she found a cute shirt that she thought I should have for NEXT Valentine's day. I smiled and said, "Let's wait and see if we're even here next year". She rolled her eyes and said, "Oh, will you stop?" But yeah, everyone seems to think I'm *worried* about it and they can't seem to understand that I'm excited (and terribly obsessed with the idea).[35]

Yet, the moral issues raised by such unbridled enthusiasm for disaster did not go unnoticed by the posters: 'So much going on in the world. It's so exciting to see prophecy unfold, and then it's so horrible to read all the sad and evil stories in the news. I'm looking up!'[36] This, more tempered, excitement holds the seeds of doubt that bothered posters who were less convinced of their own (or others') security.

Insecurity

The preceding section is a brief paraphrase of the vast majority of the content found in the sample analysed. Generally speaking, the range of emotions expressed on the website is fairly narrow, at least in regards to (in)security. This is to be expected, recalling that the website serves as a community in which faith is reinforced via the communal performance of common belief. This paradox of security amidst danger is the most notable element for scholars of critical geopolitics and critical security studies – reminding us that the 'security for whom?' axiom of critical inquiry can be turned to less progressive purposes. For instance, the environmental devastation that premillennial dispensationalists can find so comforting can have profoundly unjust outcomes for those immediately affected by it.

However, the hegemonic sense of security found in the discussion threads makes the exceptions to that monolith all the more interesting. While the security of the posters is the most theoretically interesting element to this discourse analysis, it is the occasional *in*security of the posters that provides the emotional and

---

post deleted by the administrators).

34   Post 417803 (25 January 2008).
35   Post 450113 (16 February 2008).
36   Post 396287 (12 January 2008).

embodied element that feminist scholars have often asserted is lacking in critical geopolitics. As an example, some posters felt a lack of motivation stemming from their uncertainty regarding the lack of time until the Rapture: 'Is anyone having a hard time making plans for the future? I was just getting ready to start a new biz and I can hardly do it b/c I am soooo focused on Bible prophecy unfolding every day in the news, my own relationship with Jesus, balancing that with Bible Study..and trying to create a future for myself.'[37] Similarly, this poster eschews home improvement in a way that parallels the disordering of American domestic geopolitical space in the time leading up to the Rapture:

> I'm supposed to be making plans for a total home renovation make-over which is desperately needed and which we haven't had the means to do until now. Even a year or two ago I was looking forward to it, but now it seems like a pointless time-filler, as if there isn't enough time left to make it worth doing. I have tried to fight that mind-set to no avail, instead it keeps getting stronger. The more I spend time with Jesus, the more I long to just be with Him in my Heavenly Home, not in this one. It's hard to make plans when everything seems so temporal.[38]

This poster appears to be taking refuge from a poor socioeconomic position in the Rapture: 'Your post here fits my situation exactly, and I'm turning 67 next month. No income, no pension. no retirement. I'm living on credit and putting it all in to His hands. He has never let me down and has always taken care of me.'[39] These posters illustrate how the emotional security found in the Rapture carries within it the potential for malaise and consequent economic stagnation.

The self-ascribed inactivity of the previous group of posters contrasts with the desire for the following posters to have *more* time before the Rapture. While they look forward to the Rapture and their anticipated reunion with Jesus, they worry about their friends and family who have not yet converted: 'Knots in the stomach is definitely the order of the day. Sometimes I just feel like screaming, "LORD, GET US OUTA HERE!!!" :gaah Then I remember my unsaved loved-ones, and all the others mentioned in the Prayer Requests thread, and I'm like, "NOOOO!! NOT YET, NOT YET, NOT YET!!!" :panic'.[40] This concern for family and friends can stir up complex questions of divine justice over the timing of the Rapture:

> I feel like I'm being contradictory all the time. On the one hand, I'm praying fervently for unsaved loved ones. But in the same breath, I'm asking for the Lord to return. I'm grateful that he hasn't come yet, because if he had come only two

37   Post 449702 (16 February 2008).
38   Post 449854 (16 February 2008).
39   Post 437609 (7 February 2008).
40   Post 422975 (28 January 2008).

short months ago, I'd have been left behind. The same thought dawns on me when I think of my own family and friends. They need more time.[41]

Other posters were less worried by these questions of timing and justice:

The only reason it would make sense to hope the Lord tarry a bit longer, is because of unsaved loved ones. But evenso, if He should call us up right this moment, whoever is left behind has heard and had their chance, but rejected Him. It's their own choice, and it will sadly be too late to escape. They will still have another opportunity to repent and accept Jesus during those dreadful days they'll face. God is very fair and just.[42]

The chance for unbelieving friends and family to repent during the Tribulation allows posters to feel less anxiety about leaving loved ones behind in the Rapture, but that sense of personal security is tempered by their perceived foreknowledge of the suffering to take place in the Tribulation. Hence, this woman's prayer:

Dear God, I can't wait until Jesus comes again … I am so impatient … but I have soooo many things that are bothering my heart and mind lately … At the same time, I know that my husband won't be going and I can't imagine or comprehend what he is going to have to go through … and not just him, but there are a lot of people that will be in a lot of pain … so as much as I want it now..there are people who may come to you and if you wait just a little bit longer someone's soul might be saved …[43]

Other posters felt anxiety over both the suffering that individuals would feel, and also for the destruction of temporal institutions:

If your answer to the collapse of America as we know it is just "that's just life" then you are a stronger person then me because the thought of that breaks my heart and literally makes me sick. I appreciate profoundly all the blessings we have enjoyed here in America and I love this country and the people in it very much. If having the privilege of living in the end times means that I have to watch the destruction of all I love along with family and friends then I am just saying that perhaps I would like the end times to wait awhile.[44]

This ambivalence about the Rapture and the unfolding of scripture is fascinating because it flies in the face of the 'Maranatha!' enthusiasm detailed in the earlier section. Theologically, this type of attitude should not exist as enthusiasm for the

41   Post 421186 (28 January 2008).
42   Post 428241 (31 January 2008).
43   Post 402202 (16 January 2008).
44   Post 461726 (24 February 2008).

Rapture is somewhat institutional (as reflected in the vast majority of postings). Nevertheless, a variety of human anxieties creep into play among some of the posters when considering the impending Rapture, resulting from the various ways that they relate to their belief and also the degree of strength to their faith. For instance, this poster focuses on the declining state of the world and worries that that the Rapture is too far away to spare them pain: 'We, as Christians, may very well have pain before the rapture. It is a fear of mine as well. I pray and ask God to please help to me stay true and not give in to earthly weakness.'[45] Another poster fears that he is not going to be Raptured: 'I am ready to go although I do fear I may not be ready spiritually. I still endulge and entertain the luxuries of today's society, I freely think of power and money, my lifestyle could be called questionable BUT, each day, hours on hours I think of the Lord and what is going on today.'[46]

Thus far, the anxiety discussed in this section has been rather abstract (with a few exceptions). However, many posters described embodied manifestations of their anxiety. These posters feel the affective impact of geopolitical events both cognitively and biologically, with both social and health-related consequences. The social consequences include a tendency to 'tune-out' of the news media: 'I am not one who likes anticipation. I have had to actually limit my reading of news because of the anxiety that I feel. There is so much trouble from our neighbors to our world that I ache with pain.'[47] More drastically, some posters reported trouble engaging with a world they perceive as in a downward spiral: 'This is sad to admit, but there are days when I find it hard to get out of bed. I hate this world and it's evilness.'[48]

Buttressing the fragile subject through medical/psychological intervention is, as Katz noted, prevalent in the United States.[49] Consider this poster:

> I have been feeling REALLY uneasy and at times panic stricken. I have been on anti-anxiety medication for a few years now, and at times that doesn't even seem to help. No matter where I go I see soooo much evil and madness in this world. The only time I get peace is when I am reading the Bible and praying. Other than that I always feel this pressure in my chest my mind flooded with all the things going on in this world.[50]

While this poster cites the experience of reading the Bible as the only thing that brings peace, it is unclear how much of this panic and anxiety results from the teleological world-view of geopolitical decline found in premillennial

---

45   Post 450761 (17 February 2008).
46   Post 458049 (21 February 2008).
47   Post 448956 (13 February 2008).
48   Post 448903 (15 February 2008).
49   C. Katz, 'Childhood as Spectacle: Relays of Anxiety and the Reconfiguration of the Child', *Cultural Geographies* 15/1 (2008) 5–17.
50   Rapture Ready discussion board post 438233 (7 February 2008).

dispensationalism. Nevertheless, it appears that premillennial dispensationalism does not always provide a refuge from the anxieties associated with the post-9/11 world.

Conclusions

This chapter has sought to explain the context in which an often counter-intuitive geopolitics has emerged, in which traditional notions of security have been eclipsed by an apocalyptic notion of security that finds safety in danger, security in conflict, and relief from anxiety in the study of the end of the world. The centrality of (and simultaneously, lack of) fear and anxiety to both premillennial dispensationalists' narratives of faith and self following 11 September 2001, as well as the disciplining of group boundaries, illustrates how critical actual understandings of security can differ greatly from the state-centric scale at which it is traditionally formulated. Thus, this chapter has also sought to contribute to an ongoing shift in geopolitical thought about the scale at which security should be considered. By focusing on the everyday interactions of individuals, and incorporating individuals' emotional security into geopolitical analysis, it becomes possible to account for the wide-ranging implication of the geopolitical in everyday life.

The implications of new notions of human security are manifest for the organization of political resistance to the actions of imperial governance in the US and elsewhere. While organization predicated on fostering empathy for the other has generated some positive results, it has not been sufficient to overturn the imperial consensus. What headway has been made in US public opinion has been largely based on more 'realist' considerations of the cost in American blood, treasure, and prestige. A pragmatic decision to focus on the impact of empire on peace of mind and human security of domestic populations may generate even greater returns. While this may strike some as a cynical move, catering to people's worst instincts, it is not. After all, the victims of empire are not solely 'over there'. Mediated vulnerability, securitization, and its subsequent anxiety have a cost for domestic populations as well. Future research should seek to find politically progressive ways of denaturalizing citizens' sense of geopolitical imprisonment and mobilizing these citizens to build a new post-imperial future.

Chapter 13

Common Ground? Anti-Imperialism in UK Anti-War Movements

Richard Phillips

Introduction

When the leaders of Stop the War Coalition, the organization at the heart of the anti-war campaigns in England, published a 'guide for the movement' in 2003, they did so under the title *Anti-Imperialism*.[1] This reflected the language of the anti-war movements in Britain and other countries. In Vancouver, Canada, an anti-war, anti-occupation demonstration in March 2006 was presented as fundamentally anti-imperial: 'Against the Imperialist War Drive,' read posters produced by the organizers, Mobilization against War and Occupation.[2] This was not isolated rhetoric. Many different people had been speaking about imperialism, and not only in the context of the protests against the invasion and occupation of Afghanistan and Iraq. Former colonial subjects, leaders of ex-colonies and their allies and friends in the west all found occasion to speak out against imperialism. President Robert Mugabe of Zimbabwe, for example, blaming the British (particularly their nineteenth century land grabs) for the ongoing crisis in his country;[3] and President Hugo Chavez of Venezuela accusing King Juan Carlos of Spain of an 'enduring colonial superiority complex'.[4] At the same time, others have spoken about imperialism in more nuanced or even positive terms: (then Chancellor) Gordon Brown insisting that British people should be proud of their imperial history;[5] Gideon Rachman using his column in the (London) *Financial Times* to discuss 'whether it is analytically useful to think of America as an imperial power';[6] historian Niall Ferguson lecturing students, television audiences and most recently

---

1   G. Monbiot, T. Benn, L. German, Ali, J. Corbyn and G. Galloway (eds), *Anti-Imperialism: A Guide for the Movement* (London: Bookmarks 2003).

2   <http://www.mabovancouver.org>.

3   A. Meldrum, 'Mugabe Turns Back on West and Looks East', *Guardian* (19 April 2005) 14.

4   R. Carroll and P. Hamilos, 'Royal Insult Echoes Persecution of Christ, Says Venezuelan Leader', *Guardian* (13 November 2007) 22.

5   B. Brogan, 'Let's Be Proud of the Empire, Says Brown', *Daily Mail* (14 September 2004) 26.

6   <www.ft.com/rachmanblog>.

a hedge fund on British imperialism (past) and American imperialism (present).[7] So there was nothing particularly new about the discourse of anti-imperialism in the anti-war movements. On the contrary, it was possible for activists to speak in these terms because the words they used already had some currency in a wide variety of places: from some deeply rooted predicaments (such as land disputes in Africa) to others that are or appear to be newer (the conflicts in Iraq and Afghanistan, which in reality have deep historical roots). In the war on terror, many of these disparate peoples, each with their own traditions of thinking and speaking about imperialism, were able to converge, to forge an eclectic anti-imperial movement. Anti-imperialism was the raw material from which a new, metaphorical geography of connection was forged.

As common ground for anti-war movements, anti-imperialism can be understood as a site of resistance to the war on terror. While, as other chapters in this book show, the war on terror is spatially constituted – produced in and productive of a series of new geographies of terror and security – the same can be said of resistance to this war.[8] The anti-war movements have been conceived in, advanced through and shaped by a series of strategic locations. Friends House in London, for instance, where meetings in September 2001 established Stop the War Coalition (hereafter StW), and places in which protests have been held: above all, the streets and public spaces of London, Glasgow and other cities around the world. While the timing of what its organizers accurately call Britain's biggest demonstration – 15 February 2003, effectively the eve of the invasion of Iraq – was significant in its success, its spacing was equally poignant: it passed through symbolic geographies of British imperial and military power along the Embankment and through Trafalgar Square.[9] So, too, were less tangible and bounded – virtual and connective – sites of resistance. Following earlier criticism that research on spaces of resistance[10] looked within rather than between the social and spatial contexts of political action,[11] attention is increasingly being turned to geographies of political

---

7   J. Thomas-Corr, 'The Historian Who is Helping a Hedge Fund to See the Future', *CityAM* (28 September 2007) 14; N. Ferguson, *Colossus: The Rise and Fall of the American Empire* (London: Penguin 2004).

8   UK counter terror officials belatedly recognized that the term 'war on terror' is offensive to many people and signalled a change in direction in November 2007. Ministers would no longer use this term, and would instead opt for 'less emotive language, emphasising the criminal nature of the plots and conspiracies'. R. Norton-Taylor, 'Anti-Terror Rhetoric to be Softened', *Guardian* (20 November 2007) 8.

9   F. Driver and D. Gilbert, *Imperial Cities: Landscape, Display and Identity* (Manchester: Manchester University Press 1999).

10   S. Pile and M. Keith (eds), *Geographies of Resistance* (London: Routledge 1997); J.P. Sharp, P. Routledge, C. Philo and R. Paddison (eds), *Entanglements of Power: Geographies of Domination/Resistance* (London: Routledge 2000).

11   D. Featherstone, 'Spatialities of Trans-National Resistance to Globalisation: The Maps of Grievance of the Inter-Continental Caravan', *Transactions of the Institute of British Geographers* 28/4 (2003) 404–21.

networks, notably in relation to inter-community and inter-faith coalitions[12] and transnational social and political movements[13], and from material to metaphorical or virtual spaces of cultural politics. Developing these threads, and examining anti-imperialism as a form of common ground for the anti-war movements, this chapter examines metaphorical and relational, networked spaces of resistance to the war on terror.

Alison Blunt and Cheryl McEwan suggest a two-fold agenda for postcolonial geographies, which applies equally to critical geopolitics,[14] and which addresses resistance on two levels: asking how imperial and colonial geographies are both constituted and resisted geographically; and actively intervening against these patterns of domination and subordination.[15] If we apply this challenging agenda to the geographical analysis of the contested war on terror, then we must recognize that it is not enough to locate, acknowledge and describe resistance – something the security services are busy enough doing already – we must also seek to bring the spirit of resistance into our own project. This chapter is framed by the second of the objectives set out by Blunt and McEwan, which in practice can be difficult to define, recognize and achieve, while its empirical focus is on the first, the more tangible subject of resistance to the war on terror.

Examining anti-imperialism in the UK anti-war movements, this chapter begins by examining the mobilization of anti-imperial rhetoric, examining how and why different groups of anti-war activists in the UK came to be speaking this way. Though concerned with language and with intangible spaces – 'common ground' – the discussion does not lose sight of the material geographies in which this way of thinking and speaking has been produced. It traces the series of anti-imperial traditions to a series of contrasting – variously bounded, diasporic and abstract – geographies. The next section examines the connections between these anti-imperial traditions, the ways in which they converge, not only around shared interests and projects but also the common understandings and imperatives of anti-imperialism. This convergence was, however, uneven and contested. The groups involved in the anti-war movements remained resolutely distinct. Leaders of StW claimed to be at the helm of 'Britain's biggest mass movement',[16] but in fact it remained an eclectic movement. A multitude rather than a mass,[17] this was

---

12   J. Wills, 'Campaigning for Low Paid Workers: The East London Community Organisation (TELCO) Living Wage Campaign', in W. Brown, G. Healy, E. Heery and P. Taylor (eds), *The Future of Worker Representation* (Oxford: Oxford University Press 2004) 264–83.

13   D. Featherstone, R. Phillips and J. Waters (eds), 'Special Issue: Spatialities of Transnational Networks', *Global Networks* 7/4 (2007) 383–501.

14   G. Ò Tuathail, *Critical Geopolitics* (London: Routledge 1996).

15   A. Blunt and C. McEwan, *Postcolonial Geographies* (London: Continuum 2002).

16   A. Murray and L. German, *Stop the War: The Story of Britain's Biggest Mass Movement* (London: Bookmarks 2005). Quotation from subtitle of book.

17   M. Hardt and A. Negri, *Multitude* (London: Hamish Hamilton 2005).

a network rather than a centred movement, a series of interconnected groups and protests. This limited convergence was expressed in different and specific ways, including in anti-imperial rhetoric, which different groups used and understood differently, some not at all.

This analysis draws upon interviews conducted in the autumn of 2006 and spring of 2007 by myself and two research assistants – Naima Bouteldja and Jamil Iqbal[18] – with leaders, members, supporters and critics of three groups at the heart of the UK anti-war movements, each of which brought anti-imperial traditions to their anti-war activism. Activists were not asked directly about imperialism, but in broad-ranging interviews about their anti-war experiences some mentioned this and related terms such as colonialism and empire, and were then asked to explain what they meant, also how and when they might speak in this way. Interviews were conducted with members and supporters of two of the three partners at the heart of the English movements: StW and the Muslim Association of Britain (MAB).[19] Also, the main Welsh and Scottish nationalist groups, Plaid Cymru and the Scottish Nationalist Party (SNP), which played important parts in steering and forging anti-war activism in their respective regions.

Anti-Imperialism in the UK Anti-War Movements

Martin Empson, the Chair of a London StW group and a member of the SWP, observed that 'the phrase Imperialism or the description of different countries as being imperialist nations, is something that's become commonplace in a way that it wasn't five, ten years ago'.[20] A once-fringe political vocabulary had found its way into the political mainstream, with help from figures such as Empson on the left, and others in Islamist and (Welsh and Scottish) nationalist groups, all of whom used their positions within a genuinely if briefly popular anti-war campaign to bring old rhetoric to new audiences and contexts. This section shows how the new anti-war anti-imperialism emerged out of a series of political traditions and locations.

First, socialists: though a broad coalition at first, StW was led from the left and has been dominated by the left, particularly by members of the Socialist Workers Party (SWP). Many of these activists articulated their opposition to war as a form of anti-imperialism, and in so doing they deployed a tradition of anti-imperial thinking and speaking they had been cultivating for some time. In Britain the left has a history of solidarity with colonized peoples and support for anti-colonial

---

18 Interviews are identified in this paper with details of: the interviewee's name and affiliation or position when the interview took place; the interviewer(s) (where Naima Bouteldja is NB, Jamil Iqbal JI and Richard Phillips RP); and the date on which the interview took place.

19 S. Yaqoob, 'Global and Local Echoes of the Anti-War Movement: A British Muslim Perspective', *International Socialism* 100 (2003) 39–63.

20 Martin Empson, StW, RP (12 October 2006).

struggles.[21] Tony Benn, who helped found the Movement for Colonial Freedom in 1954 to lobby for decolonization (following independence in India but in advance of that in British Africa), was to play a central role in a series of anti-war campaigns, which he presented as anti-imperial. In the most recent struggle, Benn joined others on the left including Lindsey German, John Rees, Chris Nineham of the Socialist Workers Party (SWP) and Liz Davies of the Socialist Alliance.[22] Most of the latter were too young to have had much direct experience of lobbying for the dismantling of the British Empire, but they actively inherited and remembered those movements. John Rees is the author of a book about British and American imperialism in the Middle East – *Imperialism and Resistance* – and he has referred to long histories of anti-imperialism in many of his addresses from StW platforms.[23]

Though sometimes grounded directly in anti-colonial struggles, left-wing anti-imperialism can also be traced to broader contexts, interpreted and distilled in more abstract theories. Texts on imperialism are debated on the left in discussion groups and teach-ins, in conversations between friends and in political newspapers. Martin Empson, quoted above, was one of many anti-war activists I interviewed who spoke unprompted of imperialism. Asked what he meant by the term, Empson told me he grounds his understanding in Lenin and Bukharin.[24] He and others on the anti-war left also mentioned Marxist books about imperialism such as *The New Mandarins of American Power* by the SWP stalwart Alex Callinicos (2003), and David Harvey's *The New Imperialism* (2003), as well as newspapers and magazines such as *Fight Racism! Fight Imperialism!* and the SWP's *Socialist Worker* and *Socialist Review* (see Plate 13.1). These debates have moved on in recent years to consider the arguments that imperialism has been fundamentally reconfigured. Oscar Reyes, the editor of *Red Pepper*, has observed 'one debate that regularly does the rounds' at the European Social Forum and other meetings 'is Empire versus imperialism – the Negri position versus a more conventionally Marxist one'.[25] These theoretical and ideological debates emerge from and speak to the real world – Lenin wrote his classic work on imperialism in the wake of the First World War, Hardt and Negri in the context of 1990s globalization – though their interventions were both quite abstract, cultivating metaphorical critical space general enough to encompass a great variety of experiences and politics. The ideological and theoretical differences between figures such as Callinicos and Negri were more interesting to themselves and a relatively small cognoscenti of academics and activists. Others on the left and in the anti-war movements have continued to speak of imperialism in a more general way. Rania Khan, a Respect

---

21   S. Howe, *Anticolonialism in British Politics: The Left and the End of Empire, 1918–1964* (Oxford: Clarendon Press 1993).

22   Murray and German (note 16).

23   J. Rees, *Imperialism and Resistance* (London: Routledge 2006).

24   Martin Empson, StW, RP (12 October 2006).

25   Email to Richard Phillips from Oscar Reyes (26 September 2007).

Councillor on Tower Hamlets Town Council, said simply that 'fighting against Imperialism means ... fighting powerful countries that want to impose on the poorer and the weaker'.[26] She explained that she spoke both as a socialist and also a Muslim; the latter first and foremost.

Plate 13.1 Left wing anti-imperial rhetoric in Stop the War demonstration outside Labour Party conference, Manchester, 23 September 2003
Source: Photograph by Richard Phillips.

Many British Muslims have immediate understandings of imperialism, as migrants and descendents of migrants from former British colonies including Bangladesh and Pakistan. Daud Abdullah, Deputy Secretary General of the Muslim Council of Britain, traces his understanding of this to his own experiences in Grenada, and to longer histories of the Caribbean and Africa. As he said in an interview, 'We in Latin America and the Caribbean identify with people in Africa with their struggles against imperialism and we share common pain'. He went on to say how he 'embraced Islam in Grenada in 1975', and how this helped him rebuild an identity that had been assaulted by colonialism.[27] This illustrates how anti-colonialism has

---

26   Rania Khan, Respect, RP (16 November 2006).
27   Daud Abdullah, MCB, JI (18 January 2007).

fed tangibly into Muslim identities, which have been shaped through experiences of disempowerment and feelings of powerlessness. The Muslim Brotherhood, for example, formed in Egypt in 1928 and since then 'has sought to fuse religious revival with anti-imperialism – resistance to foreign domination through the exaltation of Islam'.[28] Islamist groups across the political spectrum have echoed this, Hizb ut-Tahrir (HT) entitling its response to the UK Government's developing position on Iraq *The West's Weapons of Mass Destruction and Colonialist Foreign Policy*.[29] And Tariq Ramadan (who has been banned from entering the United States) generalizes that resistance to western colonialism has been one of the fundamental threads of political Islam, or Islamism, in the modern period.[30]

Though some British Muslims have personal experiences of colonialism, many talk about being collectively colonized, first by Europeans and more recently Americans, in what Tariq Ali calls 'a clash between native people who happen to be Muslims and the world's largest, most powerful empire'.[31] This explains why political Islam or Islamism is fundamentally anti-colonial, 'concerned', as Susan Bucks-Morss puts it, 'with challenging the hegemony of western political and cultural norms'.[32] Anti-colonial Islamism was forged in a series of colonies and former colonies, slightly differently in each, and also through a shared sense of belonging to a colonized *Ummah* with Palestine at its core: from the Crusades to the British Mandate to United States support for Israel, the colonization of Palestine is seen as representative of the relationship between western powers and Muslim nations and minorities. Daud Abdullah, who is also a founding member of the Palestine Return Centre, explained that for many Muslims personal experiences ultimately take second place to a collective consciousness, with its epicentre in Palestine.[33] Palestine is symptomatic of a two-fold grievance: against western aggression towards Muslims, and against imperialism in general. Dr Azzam Tamimi, a former President of MAB and architect of the partnership between that organization and StW, explains: 'Palestine is symptomatic of what has happened to the *Ummah* when it … came under imperial onslaught, imperialist onslaught and was cut into pieces, partitioned and puppet governments have been installed. You cannot think of what's wrong with the *Ummah* without thinking of Palestine because Palestine is at the core of this.'[34] And so, in demonstrations of solidarity

---

28   R. Leiken and S. Brooke, 'The Moderate Muslim Brotherhood', *Foreign Affairs* 86/2 (2007) 107–121.

29   Hizb ut-Tahrir Britain, *The West's Weapons of Mass Destruction and Colonialist Foreign Policy* (London: Khilafah Publications 3 November 2002).

30   Tariq Ramadan, Islamic Politics and Islamism, Seminar, University of London (10 December 2007).

31   D. Barsamian and T. Ali *Conversations with Tariq Ali: Speaking of Empire and Resistance* (New York: New Press 2005) p. 42.

32   S. Bucks-Morss, *Thinking Past Terror, Islamism and Critical Theory on the Left* (London: Verso 2003) p. 2.

33   Daud Abdullah, MCB, JI (18 January 2007).

34   Azzam Tamimi, MAB, RP/NB (17 February 2007).

with Palestinians, against Israeli occupation and military action such as its 2006 invasion of Lebanon, and more generally, Muslims frequently speak of and against imperialism (see Plate 13.2).

Plate 13.2 Anti-imperial rhetoric in demonstration against US-backed Israeli invasion of Lebanon, London, 6 August 2006
Source: Photograph by Richard Phillips.

Whereas Islamist anti-imperialism can be situated within the unbounded, disaporic geographies of the *Ummah* on the one hand, and within a preoccupation with particular colonial geographies on the other, an additional tradition of anti-imperialism can be found within much more conventionally bounded geographies: those of the disputed territories of the Celtic fringe. Nationalists and separatists oppose what they see as the English domination of Northern Ireland, Wales, Cornwall and Scotland, and make connections between Britain's domestic and foreign policies, its internal and external colonialism.[35] For Welsh and Scottish nationalists, for instance, being colonized is a fact of life and a political point of departure. Lyn Lewis Dafis, a Welsh-speaking Welsh nationalist, described Wales

---

35 M. Hechter, *Internal Colonialism: The Celtic Fringe in British National Development, 1536-1966* (London: Routledge & Kegan Paul 1975).

as 'a classic colony and probably one of the first colonies, or the first colony'.[36] He explained that Wales remains 'a colony because you know it's all set up not to be self-governing, it was set up to be part of the – the State is all set up and communication and everything is set up for the southeast of England.'[37] Those who complain of colonialism – as for instance in the slogans daubed on a dry stone wall in Anglesey (Plate 13.3) – or more specifically of internal colonialism are generally careful to distinguish it from its overseas counterparts, acknowledging that its subjects are absorbed within the colonizing society, and in some cases active agents of colonialism.[38] Nicola Fisher, chair of Stop the War in Glasgow and a former member of the SNP, described the colonization of Scotland as 'colonialism lite' (after a term coined by Michael Ignatieff: 'empire lite').[39]

Plate 13.3 'English colonists out': Anti-imperial rhetoric on stone wall in Anglesey, North Wales, August 2007

Source: Photograph by Rubinah Chowdharry.

36   Lyn Lewis Dafis, Plaid Cymru, RP (7 October 2006).

37   Ibid.

38   Hechter (note 35); G. Macdonald, 'Postcolonialism and Scottish Studies', *New Formations* 59 (2006) 116–31.

39   Nicola Fisher, Chair of StW Glasgow, RP (9 February 2007). M. Ignatieff, *Empire Lite: Nation Building in Bosnia, Kosovo, Afghanistan* (London: Vintage 2003).

Welsh and Scottish nationalism both have histories of drawing together anti-imperialism and anti-militarism. At odds with the British state, these nationalist groups have opposed the militarization that underpins it.[40] Plaid Cymru adopted a policy of opposition to all nuclear weapons in 1946,[41] and has taken a strong stand against militarism and imperialism in different forms, for example in party leader Gwynfor Evans' opposition to the Vietnam War. This tradition has continued, for example through the work of Adam Price, the Plaid MP for Carmarthen whose constant criticism of the UK's role in Iraq include initiatives to impeach Prime Minister Tony Blair, and through the SNP's longstanding opposition to nuclear weapons, championed by William Wolfe in the 1970s and by leaders and activists since.[42] SNP members have taken direct and widely publicized action against Holy Loch nuclear base near Glasgow, and the party has maintained a close relationship with Scottish CND.

Variations on a theme, these three expressions of anti-imperialism begin to illustrate how repetitive anti-imperial discourse can be. The conventional form of anti-imperialism mirrors that identified by Edward Said in its nemesis: colonial and, more specifically, Orientalist discourse.[43] Though Said's critics have shown how colonial discourse is structured by and varies with gender, sexuality and other forms of social difference – how it is more varied than Said seemed to suggest – the original thesis that highly repetitive written and spoken discourse played an important part in the production of imperialism remains persuasive and enlightening.[44] Indeed, the studies of colonial discourse that directly address its elements of repetition and performance, and link these to the production of power relations, are often those that throw most light on the production and contestation of imperialism.[45] It is in the articulation and reinforcement of claims about imperialism – for instance, in slogans and verses that are chanted and sung at demonstrations, shouted from the platforms and displayed on posters and placards at the same events – that relationships and processes are recognized, understood and sometimes resisted as imperialism. The relative uniformity of this discourse, despite the wide range of contexts to which it is applied, makes it clear that this discourse does not simply describe phenomena: it actively names them, groups them together, produces its own category or categories through them. Without deflecting attention from the materiality of the issues raised through these interventions – the land crisis in Zimbabwe, the US domination of Bolivia and

40   P. Lynch, *SNP: The History of the Scottish National Party* (Cardiff: Welsh Academic Press 2002); L. McAllister, *Plaid Cymru: The Emergence of a Political Party* (Bridgend: Seren 2001).

41   J. Ruddock, *CND Scrapbook* (London: Optima 1978).

42   William Wolfe, former Leader of SNP, RP (2 December 2006).

43   E. Said, *Orientalism* (London: Penguin 1978).

44   R. Lewis, *Gendering Orientalism: Race, Femininity and Representation* (London: Routledge 1996).

45   D. Gregory, *The Colonial Present* (Oxford: Blackwell 2004).

Venezuela, the wars in Afghanistan and Iraq, and so on – the common language makes it clear that there is something more going on here: an identification of individual experiences with something more general, and an identification with others that share these experiences in some way. This raises questions, addressed in the next section, about how parallel traditions of anti-imperialism converged, to the extent that they did, producing common ground and using it to drive the anti-war movements forward.

Geographies of Connection

Remaining differentiated and decentred, the anti-war movements did not subsume so much as bridge differences between their constituent groups. Many activists – including Welsh and Scottish nationalists and some Muslim groups – insisted on a mixture of local and national action, rather than mere participation in a movement centred in London. And so, despite some spectacular demonstrations in the capital, protests were organized and took place in and many smaller and quieter places: a stall set up outside the Welsh language bookshop in Aberystwyth, demonstrations in the main street of Ullapool in the Scottish Highlands and a local park off Brick Lane in London. So while many activists and communities supported some of the London demonstrations, they did not pour all their energy into them, or subordinate themselves to the London-based StW. Much to the irritation of those who would like to see the movements more centralized and homogeneous – notably the journalist Nick Cohen, who complains of their 'communalism'[46] – activists have continued to identify with distinct groups: Scots and Welsh distinguishing themselves from English activists, Muslims from non-Muslims, and so on. Salma Yaqoob quotes a young activist to suggest that the movements bridged differences between Muslims and others, while keeping those differences clear and intact: 'I led 500–600 students on a walkout from my college through the streets of Birmingham with a feeling of unity and peace … It is amazingly reassuring to see Muslims and non-Muslims uniting on this issue,' said then 17-year-old Umbreen Hussain.[47] Connections between activists and demonstrations in England and those in Wales and Scotland, and still others in Rome, Ankara, Sydney, Seoul and across the world, underline the ways in which the movements were able to cross borders and cut across – but not eliminate – differences. Forging and negotiating mutual connections and common ground on which they were able to meet, activists developed expanding networks and conversations, which provided their movement with a dynamic.

---

46   N. Cohen, *What's Left? How Liberals Lost Their Way* (London: Fourth Estate 2007).

47   Quoted by Yaqoob (note 19) p. 43.

If the anti-war movements were a politics of what Hardt and Negri[48] call the 'multitude' rather than the masses, this was true not only of their social composition and material geographies, but also their metaphorical spaces: the common ground in which activists came together to forge collective projects, which was never an absolute space-in-itself, but rather an open-ended geography of connection. While distinct anti-imperial traditions can be traced to particular geographies – from the disaporic spaces of the *Ummah* to the bounded territories of Welsh and Scottish nationalism and the more general, abstract spaces of the left – the convergence between them can too: in this case to relational spaces. The following paragraphs trace material practices through which these connections take place, then goes on to assess the extent to which this has produced genuine understanding and political common ground, in which disparate groups have been able to communicate with each other and develop common projects. This, in Susan Bucks-Morss's more general terms, means 'producing solidarity beyond and across the discursive terrains that determine our present collective identifications'.[49]

The extension of ideas about imperialism and colonialism from one set of discursive terrains to another, and from classic colonies to variations on the theme such as 'internal' colonialism among territorial or social minorities and intangible forms of contemporary domination and subordination, effectively 'stretches' these ideas, applying them to contexts where their meanings are substantially different. This is both a metaphorical and also a material stretch: from one subject to another, but also from one place to another.[50] Stretching ideas and language across material geographies involves a series of different practices: writing, printing, publishing and reading texts, and producing, communicating and consuming (hearing and/or watching) broadcasts and other immaterial texts. Stretching a metaphorical space is therefore in many ways a material practice, produced in and through a series of tangible spaces and materially grounded representational practices, which in turn stretch ideas and realities of imperial power and anti-imperial resistance across great geographical and social distances.[51]

To begin to speak of the material practices that stretch discourses is to raise questions about the concrete ways in which anti-imperialisms have been extended and converged in the anti-war movements. One of these mechanisms is through shared genealogies, activists drawing upon some of the same sources: language, understandings, and texts. Though each of the groups considered here has its own specific influences – SWP activists were the only ones to mention reading

---

48   M. Hardt and A. Negri, *Empire* (Cambridge, MA: Harvard University Press 2000); Hardt and Negri (note 17).

49   Bucks-Morss (note 32) p. 4.

50   R. Jones and R. Phillips, 'Unsettling Geographical Horizons: Premodern and Non-European Imperialism', *Annals of Association of American Geographers* 95/1 (2005) 141–61.

51   M. Ogborn, *Indian Ink: Script and Print in the Making of the East India Company* (Chicago and London: Chicago University Press 2007).

certain theorists such as Alex Callinicos,[52] for example, and Welsh translations of the biographies of Mahatma Gandhi and Martin Luther King could only be cited by activists in Welsh-speaking Wales[53] – the influence of other writers was much wider. Writers such as Noam Chomsky, Robert Fisk, John Pilger and Arundhati Roy were cited by members of each of the three groups interviewed for this project, as were historical figures such as Gandhi.[54] In this respect, though the different groups claim their own traditions and influences, they also take inspiration from some shared figures and texts.

A second way in which these groups and traditions are intertwined is through overlaps and linkages between them: individuals who are members of more than one group. For example, some Muslims living in Scotland identify as Scottish nationalists as well as Muslims, as Peter Hopkins has shown in his work on young Scottish Muslims.[55] Osama Saeed[56] identifies both a Muslim – a spokesperson for MAB in Scotland – and a Scottish nationalist, selected by the SNP to represent a Glasgow constituency in the Scottish Parliament. There is also overlap between each of these groups and the left. For example: Asad Khan,[57] the coordinator of StW in Bury, and Noreen Fatima,[58] chair of StW's Muslim Network, are both Muslims and socialists; Leanne Wood,[59] Plaid Cymru Assembly Member, is both a socialist and a Welsh nationalist.

These groups are also interconnected through perceived affinities. With its essentially Trotskyite SWP leadership, StW is fundamentally internationalist and rooted in solidarity with working classes and colonized peoples wherever they may be; Muslim anti-imperialism is rooted in an identification with the *Ummah* and with other subjects of western colonialism; and Welsh and Scottish nationalism is energized by identification with other peoples who are colonized within nations. Illustrating this point, Plaid Cymru MEP Jill Evans said, 'I think that the Welsh people certainly have an empathy with oppressed people such as in the case of Palestine because we're a small nation and because there have been long struggles to get status for the Welsh language for instance. And it's very different, it's a completely different experience and you can't compare the two things I don't think'.[60] Other Welsh nationalists spoke of their affinity and ability

---

52   For example, Martin Empson, StW, RP (12 October 2006).

53   Richard Owen, Plaid Cymru, RP (7 October 2006); Dafydd Morgan Lewis, Cymdeithas yr Iaith Gymraeg (Welsh Language Society) and Plaid Cymru, RP (7 October 2006).

54   Mark Holt, StW, RP (25 October 2007); Roddy Slorach, SWP, RP (27 October 2006).

55   P. Hopkins, 'Young, Male and Scottish: A Portrait of Kabir', paper presented at Annual Meeting of Association of American Geographers, San Francisco (2007).

56   Osama Saeed, MAB, SNP, NB (23 February 2007).

57   Asad Khan, StW Bury, JI (22 June 2007).

58   Noreen Fatima, StW Muslim Network, NB (12 April 2007).

59   Leanne Wood, Plaid Cymru Member of Welsh Assembly, RP (2 November 2006).

60   Jill Evans, Plaid Cymru MEP, Chair of CND Cymru, RP (22 November 2006).

to 'empathize'[61] with minority and would-be nations and peoples who 'are feeling that they are being occupied by imperial forces like Britain and America'.[62] They mentioned nations such as Scotland and Catalonia, and oppressed peoples from African Americans to Guyanese, and from Kurds to Marsh Arabs – both of whom have been regularly received on the platform of Plaid Cymru conferences. For Dafis, 'civil rights in America is a great inspiration',[63] and Leanne Wood, Plaid Cymru Member in the Welsh Assembly, says that though she is white and Welsh, she feels a connection with certain African American women. 'I'm a major fan of Alice Walker and Toni Morrison and all those black women writing in the southern States,' she said. 'I just really felt that I had a massive connection with these women from being a working class woman in an industrialized area.'[64] Similarly, people in Scotland spoke of their affinities with small northern and/or postcolonial European nations including Ireland and Norway. For Niamh O'Toole, a StW activist and member of the SWP living in Glasgow, 'the difference between Scotland and Ireland' is something 'you could put a hair between'.[65] And for their part, Muslims identified not only with their brothers and sisters in the *Ummah*, but also with others they perceive to be pitted against and bullied by the British and other western states, notably the Irish.

While many people simply perceived affinities, others acted upon them, forging links with others by starting conversations and forging relationships. Welsh-speaking anti-war activist Sian ap Gwynfor described a 'shock of recognition' in Guyana (formerly British Guyana), where she witnessed 'festering wounds of imperialism'.[66] Welsh and Scottish nationalists spoke of cooperation at Westminster and in Brussels, as well as through the Inter Isles Forum: a youth wing project bringing together members of the SNP, Plaid Cymru, Sinn Fein, the SDLP, DUP and other political parties from across the British Isles and Ireland.[67] Connecting the experiences of Irish immigrants in the 1970s with those of Muslims today, seeing how both have lived under clouds of suspicion and surveillance, ex-Guantánamo detainee Moazzam Begg embarked upon a visit to Ireland, an attempt to build bridges and common politics. He explained how he had 'stepped out towards and reached out to' other peoples, non-Muslims:

---

61   Iwan ap Dafydd, Plaid Cymru, Aberystwyth, RP (6 October 2006).
62   Bethan Maeve Jenkins, Plaid Cymru, RP (1 November 2006).
63   Lyn Lewis Dafis, Plaid Cymru, RP (7 October 2006).
64   Leanne Wood, Plaid Cymru, RP (2 November 2006).
65   Niamh O'Toole, StW, SWP, RP (24 February 2007).
66   Sian ap Gwynfor, Peace Activist, RP (25 November 2006).
67   Gareth Finn, SNP-CND, RP (2 December 2006); Bethan Maeve Jenkins, Plaid Cymru, RP (1 November 2006).

Rights are being usurped around the world ... Even as a practicing Muslim I still identified with, for example, Nelson Mandela, still identified with the Irish or the Catholics in Northern Ireland and places like that.[68]

One of the common threads to all this cross-identification is a shared anti-imperialism and in some cases a shared (but distinct) experience of being colonized. For HT member Nazmul Hoque, this established a bond between groups that were, as he put it, 'totally night and day different' in many ways. Hoque said that Islamists and socialists were brought together by a common enemy: both had 'struggled with the west, with the ideas of capitalism' and with 'colonization, imperialism, and the slave trade' – 'there was a commonality in that sense'.[69] Asked what he meant by the term imperialism, he said 'we don't have an all unique definition as Muslims or from Hizb ut-Tahrir. That definition of imperialism is a common definition'.[70] For Hoque, a member of an organization that has preferred not to work directly with StW or participate in the mainstream anti-war movements (though it has belatedly moved in this direction), anti-imperialism was a rare form of common ground, something linking Muslims and others in a common struggle.

While anti-imperialism was not enough to bring disparate activists together – their opposition to more tangible expressions of the war on terror did that – it was nevertheless a vehicle for bringing them together. Imperialism, as Stephen Slemon has argued, has room for unity and difference, since it 'is by definition transhistorical and unspecific, and it is used in relation to very different kinds of historical oppression and economic control'.[71] The effect of incorporating a range of experiences and processes within a common theory and language need not elide the differences between them, but on the contrary may provide a conceptual framework for analysing those differences. Its articulation of unity and difference helped anti-war activists with different backgrounds feel a bit more united than they might otherwise have done.

Distances remained between these groups and communities, mainly because of their different agendas for the anti-war movements and their attachment to different cultural and political traditions. The rise and fall of the partnership between StW and MAB, which was forged in April 2002 and lasted until January 2005,[72] was shaped by the limited overlap between their agenda: StW was primarily concerned with Afghanistan and Iraq, while MAB attached equal importance to Palestine; and by the ability of secular socialists to work an often conservative religious community and vice versa. Their partial convergence was also reflected in their

---

68   Moazzam Begg, Ex-Guantánamo Detainee, RP/NB (21 May 2007).

69   Nazmul Hoque, HT, JI (19 February 2007).

70   Nazmul Hoque, HT, JI (19 February 2007).

71   S. Slemon, 'Unsettling the Empire: Resistance Theory for the "Second World"', *World Literature in English* 30/2 (1990) 30–41.

72   For detailed account of this, see: Richard Phillips 'Working Together? The Muslim Association of Britain and Stop the War Coalition', *Race and Class* forthcoming.

attempts to communicate and to find common ground. Like many other things, anti-imperialism brought them together, but only to a point.

The more general contestation of anti-imperialism was evident from the first days of the anti-war struggle. In the formative anti-war meetings that took place in London and other cities after 11 September 2001, debates took place about what its founding principles should be. Motions were proposed and put to the vote. Some of these – on opposition to war, racism and the erosion of civil liberties – were passed, while others including an 'attempt to demand that all those who oppose the war also oppose imperialism' were not.[73] This did not stop leaders of the new Coalition from speaking about imperialism, and publishing their 'guide to the movement' under the heading *Anti-Imperialism*.[74] Common ground, they decided, was always up for grabs. As the movements unfolded, rose and then apparently fell, anti-imperialism formed a central but contested strand of their rhetoric. Those who spoke about and against imperialism, or heard others doing so, had mixed feelings about this rhetoric. On the one hand, different groups and communities retain their own specific understandings of imperialism, Welsh and Scottish experiences of 'colonialism lite' remaining radically different from their more brutal counterparts in Palestine and Iraq. There is, to borrow a phrase from Winston Churchill, a danger of socialists, Islamists, nationalist and other anti-imperialists being 'divided by a common language'. And on the other hand, understandings of and attitudes towards imperialism vary within these and other groups. For some people, this way of thinking and speaking is powerful and pertinent, but for others it is too broad a term, or too intellectual, abstract, simplistic, jargonistic, sloganistic, or distracting. Glyn Robbins, an anti-war activist in London who chaired a local group of Respect, the political party that formed out the anti-war coalition between socialists and Islamists, summarized these more general concerns about anti-imperialism. 'If the first thing we said was about imperialism,' he said, 'then I think we would potentially be cutting ourselves off from a number of people who maybe don't define themselves ... as primarily opposing imperialism' – people whose politics are more conventional and even conservative.[75] This language may have proliferated, but it was not always productive, and would not always draw people together, either within or between anti-war communities.

Conclusion

This chapter has shown how anti-imperial common ground within the anti-war movements crystallized out of overlapping and intersecting anti-imperial traditions, operating on different geographical scales, from action against alleged internal colonialism to national and trans-national resistance to colonial occupations. The

---

73   Murray and German (note 16) p. 49.
74   Monbiot et al. (note 1).
75   Glyn Robbins, Respect, RP (4 October 2006).

conclusion reflects on what if anything this has achieved, and what its significance is for understandings of the spatiality of power and resistance, with particular reference to the war on terror and more generally to imperialism.

On the surface, the anti-war movements have failed, and by implication their political and rhetorical strategies have failed with them. Certainly, they have failed to stop the invasions and then occupations of Afghanistan and Iraq, and seem to have done little to arrest the erosion of civil liberties and escalation of Islamophobia that have marked and marred the war on terror closer to home. And the movements themselves, both in the UK and internationally, have withered in recent years, even while the wars and occupations have continued.[76] But within these movements it is still possible to find grounds for hope, not least within the relationships they have engendered between previously disparate groups. Working together has left a legacy of understandings, friendships, relationships, or, to use a more formal social scientific term, bridging capital.[77] Muslims may be a little closer to people of other faiths or none, Welsh pacifists to Scottish anti-nuclear activists, and so on. These groups were brought together mainly by their shared opposition to the war on terror, of course, but they needed to be able to communicate with each other and cultivate a sense of togetherness for this to be possible, and ideas they could share – about imperialism, for instance – helped activists to feel more united, less 'night and day different' (as HT member Nazmul Hoque put it) at least. One reason this is encouraging is simply that it has brought different groups together. Collaboration between StW and MAB, for instance, does not necessarily speak of collaboration between wider socialist and Islamist communities, since the latter are very much wider and more diverse, but it nevertheless points suggestively in that direction, speaking of connection between at influential sections of these wider communities.

The important thing is not just that very different groups are now speaking about imperialism, but that they are engaged in a common debate, at least potentially communicating with each other, something arguably crucial to seeing beyond the war on terror, to envisaging collaborative pro-active politics and more hopeful futures. If in a sense it is good that different groups have been able to communicate, it is better that they have chosen this particular subject as the basis for their communication. Though anything might have done – families, coffee, cheese! – the conversation about imperialism is better than this, because it offers political direction. Speaking of imperialism, activists are able not only to speak to each other, but to develop a collective sense of what they are against: the sometimes apparently unstoppable 'Empire' of American militarism and global capitalism; and how, in being against it, they are together.

---

76   A. Cockburn, 'Whatever Happened to the Anti-War Movement?', *New Left Review* 46/July-August (2007) 29–38.

77   R. Putnam, *Bowling Alone: The Collapse and Revival of American Community* (New York: Touchstone 2000).

To speak of bridging distances and producing common ground is, of course, to use spatial metaphors about resistance, and to revisit questions posed at the beginning of this chapter about the significance of space – material and metaphorical – for resistance to the war on terror. This chapter has focussed upon a metaphorical, generic geography of resistance but it has also shown how this is grounded in and distilled from a series of material geographies: a range of concrete colonial experiences. This contributes to a broader understanding of how these material and metaphorical, bounded and relational spaces shaped not only the war on terror but also resistance to it, not only the constitution of but also resistance to contemporary imperialism. And this, in turn, develops key strands of the two-fold agenda for postcolonial geographies (and critical geopolitics), defined by Blunt and McEwan as addressing the production and contestation of imperial power, and itself intervening against imperialism.[78] These two purposes do not always go together, for there is nothing automatically critical or postcolonial about documenting and analysing resistance, as we can see from the range of people who do this: from postcolonial critics on the one hand to colonial police forces and counter-insurgency teams on the other. Nor is there anything automatically anti-imperial about naming imperialism, as interventions on the subject by neoconservative and establishment figures such as Richard Haas and Niall Ferguson have shown. But while there is no formula for committed research, which intervenes rather than describes, writing against rather than merely about the war on terror and other forms of domination and subordination, it is nevertheless productive to set this as an ideal. It is ultimately the inspiration for and measure of chapters and books such as this.

Acknowledgements

The research for this article was supported by Economic and Social Research Council (ESRC) Research Grant RES-000-22-1785. This paper draws upon research conducted by the author and two Research Assistants: Naima Bouteldja and Jamil Iqbal.

---

78   Blunt and McEwan (note 15).

Chapter 14

Art and the Geopolitical: Remapping Security at *Green Zone/Red Zone*

Alan Ingram

There will be some things that people will see. There will be some things that people won't see. And life goes on. (Donald Rumsfeld, 12 October 2001 Department of Defense news briefing)[1]

I employ the term [situated ignorance] because, even in this age of wireless communications and the Internet, the collective or individual encountering of discourse is always an embodied practice, occurring at dispersed sites whose particular nexus of power relations and other invisible and visible geographies continue to matter. (Allan Pred, 2005)[2]

Introduction

Much research has addressed the respatialization of the globe under the schematic imaginative geographies of the war on terror: our space/their space; homeland/target; core/gap/seam; border; black site; camp; battlespace; free-fire zone.[3] In keeping with the aims of critical geopolitics, this work has not restricted its focus to the realms of formal theory, practical policy making and mainstream media but has also addressed various elements of popular culture, including film, cartoons, comics and videogames.[4] This chapter suggests that it is worth broadening the

---

1   H. Seely, 'The Poetry of D.H. Rumsfeld' (April 2006) at <http://www.slate.com/id/2081042/>.

2   A. Pred, 'Situated Ignorance and State Terrorism: Silences, WMD, Collective Amnesia, and the Manufacture of Fear', in D. Gregory and A. Pred (eds), *Violent Geographies: Fear, Terror and Political Violence* (New York: Routledge 2005) 363–84, p. 365.

3   D. Gregory, *The Colonial Present* (Oxford: Blackwell 2004); Gregory and Pred (note 2).

4   For example, on film: K. Dodds, '"Have You Seen Any Good Films Lately?" Geopolitics, International Relations and Film', *Geography Compass* 2/2 (2008); M. Power and A. Crampton, 'Reel Geopolitics: Cinemato-graphing Political Space', *Geopolitics* 10/2 (2005) 193–203; on cartoons: K. Dodds, 'Steve Bell's Eye: Cartoons, Popular Geopolitics and the War on Terror', *Security Dialogue* 38/2 (2007) 157–77; on comics: J. Dittmer, 'The Tyranny of the Serial: Popular Geopolitics, the Nation and Comic Book Discourse',

compass of critical geopolitical reflection still further to include certain kinds of art practices.[5]

One reason for this is to take account of the upsurge in artistic interventions addressing contemporary geopolitics. Though art, galleries and museums have always been (geo)political,[6] the war on terror and contemporary securitizations seem to be fuelling an upsurge in artistic responses within the West but also beyond it, and along diasporic and intercultural trajectories that disrupt such coordinates. These responses are taking place both within and beyond conventional spaces of display, and stretch to practices such as graffiti, street art, installation and site-specific interventions.[7] While the risks of reappropriation and over-celebration of such practices are ever-present, this chapter suggests that the depth and breadth of contemporary artistic interventions warrants greater consideration by scholars of geopolitics and other geographers interested in the spatialities of identity, power and security.

Though relationships between art and the war on terror have received some consideration in international relations and security studies, these disciplines have not dwelt on their spatial dimensions.[8] However, spatiality is a key (but sometimes underappreciated) field in which the war on terror has both been propagated and resisted by a wide variety of artists and curators. Much of this is to do with the control of visuality and the geopolitical gaze.[9] If, as indicated by Donald Rumsfeld, the war on terror has been premised upon a dominant regime of visibility and

---

Antipode 39/2 (2007) 247–68; on video-games: M. Power 'Digitized Virtuosity: Video War Games and Post 9/11 Cyber-deterrence', *Security Dialogue* 38/2 (2007) 271–88.

5 Here the task of defining art is accomplished by dodging it: the term is taken to refer to aesthetic productions that are made and received as such, that is, subject to the discursive limits that enable and constrain the naming of things in this way. It therefore includes things often considered to be 'art' in only a marginal sense (e.g. graffiti) or even negatively (e.g. certain kinds of installation or things commonly subject to statements such as, 'is it really art?' or 'there is no way that this is art'). The term art practices encompasses the making and display of art works and the curatorial process.

6 T. Luke, Shows of Force: Power, Politics, and Ideology in Art Exhibitions (London: Duke University Press 1992); T. Luke, Museum Politics: Power Plays at the Exhibition (Minneapolis: University of Minnesota Press 2002).

7 F. Gavin, *Street Renegades: New Underground Art* (London: Laurence King 2007); P. Kennard, 'Art Attack', *New Statesman* (17 January 2008) at <http://www.newstatesman.com/200801170028>; E. Mathieson and X. Tàpies, *Street Art and the War on* Terror (London: Rebellion Books 2007). I also direct the reader to my blog at <http://artxgeopolitics.wordpress.com>, which I began in an attempt to connect and track artistic responses to contemporary issues of geopolitics and security.

8 See R. Bleiker, 'Art after 9/11', *Alternatives* 31/1 (2006) 77–99 and the special issue of *Security Dialogue* (38/2, 1997) on securitization, militarization and visual culture in the worlds of post-9/11; J. Der Derian, 'The Art of War after 911' at <http://www.watsoninstitute.org/news_detail.cfm?id=52>.

9 G. Ó Tuathail, *Critical Geopolitics: The Politics of Writing Global Space* (London: Routledge 1996).

control, then part of the political potential of art is its ability to place within the field of vision things that are not meant to be there. This second purpose of this chapter therefore is to show how reflection on the geographies of art practices can enlarge the sense in which they can be considered political, and that art practices that are rich in spatiality have particularly powerful political valences.

This approach to geopolitical discourse (like that which focuses on its intertextuality) has been around for something like two decades. Yet, as Johanna Drucker argues,[10] such interpretive strategies have struggled to cope with contemporary forms of power and violence. In response she proposes a strategy of refamiliarization that seeks to recover the webs of interconnection through which the world is made and to renovate ideas of accountability and responsibility:

> The task of refamiliarization is to show that what *is* is *not* entirely simulacral, but connected to the lived experience of persons and peoples, organic beings, within cultural, political, and vulnerable ecological spheres. This reconceptualizes the work that we imagine interpretation can do. The goal is not instrumentalized activism, but recognition of the rich complexities and dynamic contradictions of generative processes.[11]

Drucker's approach, I suggest, has close affinities with developments in critical geopolitics, where the initial methods adopted towards the gaze and discourse have also been subjected to critique. Proposals have been made (notably, but not only, by feminist geographers) to adopt more embodied, performative and relational conceptions of space, particularly as they enable an engagement with bodily and emotional experience and disturb distinctions between the global and local, or the geopolitical and everyday.[12] While Drucker draws something of a distance from 'instrumentalized activism', I adopt a more engaged approach, and one that is about dialogue with others as much as the interpretation of the critic. With this in mind, this chapter is intended as an illustration of how art practices

---

10   J. Drucker, 'Making Space: Image Events in an Extreme State', *Cultural Politics* 4/1 (2008) 25–45.

11   Ibid., p. 30.

12   L. Dowler and J. Sharp. 'A Feminist Geopolitics?', *Space and Polity* 53 (2001) 165–76; J. Hyndman, 'Beyond Either/Or: A Feminist Analysis of September 11th', *ACME* 2/1 (2003) 2–13; J. Hyndman, 'Mind the Gap: Bridging Feminist and Political Geography Through Geopolitics', *Political Geography* 23/3 (2004) 307–22; R. Pain and S. Smith, *Fear: Critical Geopolitics and Everyday Life* (Aldershot: Ashgate 2008); R. Pain, 'Globalized Fear? Towards an Emotional Geopolitics', forthcoming in *Progress in Human Geography*; J. Sharp, 'Remasculinizing Geo-politics? Comments on Gearoid O'Tuathail's *Critical Geopolitics*', *Political Geography* 19/3 (2000) 361–64; N. Thrift, 'It's the Little Things', in K. Dodds and D. Atkinson (eds), *Geopolitical Traditions: A Century of Geopolitical Thought* (London: Routledge 2000) 380–7. There are also affinities with the concept of conscientization discussed by Rachel Pain (forthcoming).

can be interpreted as geographical counterpositions to the situated ignorance of which Allen Pred wrote.

A further premise of the chapter is that there is in fact a convergence between the ways in which a number of artists and curators, on the one hand, and critical geopolitics, on the other, have been addressing epistemological concerns such as these, and that mutual reflection might help to foster and deepen dialogue between these two projects. However, if such dialogue is to proceed very far it must be reflexive and acknowledge the geographies of power which enable and structure it. These geographies are diverse, but this chapter addresses in particular those associated with orientalism. Though it is by no means a perfect solution, my approach (inspired in part by certain artistic and curatorial strategies), is to adopt a form of contrapuntal analysis of the kind suggested by Edward Said in an attempt to draw out connections between people and places across the dividing lines of violence and security.

Following further discussion of conceptual and methodological issues, the chapter considers as its main empirical focus the exhibition *Green Zone/Red Zone* that took place from 20 October 2007 to 31 January 2008 at Gemak, a new cultural institution located in the Hague that aims to present a 'challenging blend of contemporary art and politics'.[13] The chapter argues that *Green Zone/Red Zone* represents an excellent example of the convergence of concerns between art and critical geopolitics, and seeks to uncover a number of the geographies at work there.

Green Zone/Red Zone addressed the 'division or involuntary fragmentation of cities like Baghdad under the pretext of improving security' and is particularly interesting, I argue, for a number of reasons. Discursively, it was an explicit attempt to go beyond the limits that structure Western experiences of geography, terror and security. It also sought to disrupt the dominant regimes of visibility and control through which the war on terror has been forwarded. Displaying works of varying form and nature, it confronted visitors with a range of embodied and performative representations and experiences. And it brought together a wide variety of situated knowledges from many places caught up in the war on terror and contemporary securitizations, offering particularly rich potential for geographical counterpositions.

In order to illustrate this, I discuss the curatorial strategy of Gemak itself, survey some of the contents of *Green Zone/Red Zone*, and consider how those contents relate to the Hague as an urban space, a special site within international relations, and part of the geopolitical West. This discussion draws on my own visit to the exhibition and conversations with Robert Kluijver, curator of Gemak, and three of the artists (Rashad Selim, Peter Kennard and Cat Picton-Phillipps) who contributed works, which I then consider in more detail. I consider a wide range of works rather than a smaller selection because it is the very breadth and diversity of geographical connections and resonances in play at *Green Zone/Red Zone* that I want to highlight. Overall the chapter argues that if it is indeed axiomatic that

---

13   *Green Zone/Red Zone* flyer.

challenging simplistic formulations of space is 'a necessary part of resisting war as the dominant social relation of our times',[14] then strategies such as those in evidence at *Green Zone/Red Zone* are worthy of further consideration by geographers, and those concerned with critical geopolitics in particular.

Critical Geopolitics and Cultural Criticism

In a recent essay, Johanna Drucker reflects on the questions of what aesthetic acts are and what they can do in a culture where imaginative art appears to have been compromised by the abuse of images (as mechanisms of domination and accumulation) and by images of abuse (from, as she mentions, Rwanda to, we might easily suggest, Abu Ghraib). She suggests that in this context prevailing cultural strategies (which might seek to shock, to reveal 'the real' behind the surface of events, or to deny the concept altogether) are no longer viable and instead proposes *refamiliarization*, a rethinking of aesthetic images as events within the relations, processes and ideologies through which they come into being and within which people are also situated. Refamiliarization (which she also discerns in the work of a selection of key figures in fine art) 'introduces points of connection, associations of reference, lines of accountability, responsibility, engagement'.[15] It also seeks to create 'room for play in a bureaucratically managed universe': 'In that space of play, we can imagine – not "otherwise" but instead – into recognition, awareness of our place within ideological conditions.'[16]

Though the invocation of a unitary 'we' is invariably problematic, this approach shares things in common with wider currents in geography and social theory, and I suggest that it is a useful analytic through which to consider artistic responses to the war on terror. However, it is one that, if we are to address the geographies involved, needs to be more decisively spatialized and more reflexive in orientation. A key dimension of this is epistemological. The situated and partial (though still not wholly relativistic) nature of knowledge and science are now quite well established in parts of geography and other disciplines. As Rogoff suggests,[17] in this regard geography can be thought of as both condition to be investigated and methodological strategy, in that one way to gain a better understanding of the world is to engage in dialogue with a wider range of places, positionalities and relationships through which it is experienced and constituted, recognizing that imaginative geographies and lived experiences are heavily implicated in asymmetric power relations. Such a task is by no means straightforward or without

---

14   S. Dalby, 'The Pentagon's New Imperial Cartography', in Gregory and Pred (note 2) 295–308, p. 306.

15   Drucker (note 10) p. 43.

16   Ibid., p. 42.

17   I. Rogoff, *Terra Infirma: Geography's Visual Culture* (London: Routledge 2000).

pitfalls, but it implies a more honest position than the panopticism and other god-tricks bequeathed by the history of geopolitics.[18]

Refamiliarization might also be an appropriate strategy at this point in the war on terror. Several years in, many of its workings are now in plain light; many of the key drivers are well known and have been the subject of extensive theorization. Each new announcement, discovery or event seems hardly to add to our understanding: each missile strike from a Predator drone; each suicide bombing; 'revelations' about waterboarding or the use of Diego Garcia (another sign of British state accomplicity)[19] in the rendition, detention and torture system; and reports of the inordinate profits accruing to the military-security-industrial complex. Though shocking, these are hardly surprises: as we know this is all part of the 'new normal'. Moreover, as Debrix suggests in his discussion of 'sublime spectatorship', appalling images of war have routinely been presented and rationalized in US media and popular culture in ways that uphold prevailing ideological positions about the essential humanity of what the US is doing.[20] What might be useful are new ways to present and experience the connections and relations between the different parts of this landscape; how it fits together, where it comes apart, how people are differently located within it. At the same time, refamiliarization is not incompatible with the idea of curiosity: 'the optimism of finding out something one had not known or been able to conceive of before'; indeed, refamiliarization might expand 'the realm of the known'.[21]

It is important to stress the performative dimension to this: I consider artworks not as 'comments' 'on' or 'about' geopolitics (a kind of 'spectatorship')[22] or necessarily as rallying banners for political mobilization conventionally understood (though they may be that), but more broadly as part of the geopolitical dynamic itself: as artefacts produced by people as part of their situated and embodied experience of power projection, capitalist globalization, spectacular and covert political violence, rebordering and new strategies of surveillance and security. They are also worthy of consideration as interventions back into those processes and landscapes.[23]

This focus also raises the question of what Fraser Macdonald has termed observant practices, of 'what it means to see and how geopolitical power is exercised through the experience of sights and spectacles' but also of how sight

---

18   Hyndman, 'Beyond Either/Or' (note 12) p. 6.

19   A. Danchev, 'Accomplicity: Britian, Torture and Terror', *British Journal of Politics and International Relations* 8/4 (2006) 587–601.

20   F. Debrix, 'The Sublime Spectatorship of War: The Erasure of the Event in America's Politics of Terror and Aesthetics of Violence', *Millennium: Journal of International Studies* 34/3 (2006) 767–91.

21   Rogoff (note 17) p. 33.

22   Ibid., pp. 33–35.

23   See the discussion in R. Pain and S. Smith, 'Fear: Critical Geopolitics and Everyday Life', in Pain and Smith (note 12).

and seeing are connected to other dimensions of embodied experience.[24] There is considerable scope to explore this issue further in relation to the kinds of art practices considered here. However, for the purposes of this chapter, and though it is clearly relevant to the way that I have responded to *Green Zone/Red Zone*, the precise nature of connections between observant practices and the embodied experience of imaginative geographies remains largely implicit. The chapter remains primarily an exploration of *Green Zone/Red Zone* based on my own experience and engagement with those involved in its production rather than an analysis of how other people experienced or responded to it.

To be sure, there are risks in an investigation along these lines. Reflecting on certain trends within the art world, Drucker states,

> Conspicuously offered as works of "opposition" or "resistance", high-gloss products are promoted for their ability to "critique" the culture industry or "transgress" the symbolic discourses of power. Claims to critical enlightenment are supported by an industry – academic as well as commercial – thriving within self-regulating and self-replicating discursive systems. Whether these claims have any credibility outside these protected domains is an open (and familiar) question.[25]

Furthermore, as WJT Mitchell observes, 'We as critics may want pictures to be stronger than they actually are in order to give ourselves a sense of power in opposing, exposing or praising them.'[26] Similarly, I would agree with Lisle that

> artistic dissent is not a special kind of dissent – artists do not stand above the fray and watch (and paint, and draw, and sculpt, and sing, and write) as we mere mortals scrabble around in the detritus of politics trying to resist the homogeneous "orthodoxy" or "conventional wisdom".[27]

Art, then, is not 'magical politics' and neither its autonomy nor its efficacy can be assumed; criticism (whether aesthetic or academic) can easily become spectacle and is susceptible to commodification and reappropriation.[28] However, I proceed in the belief that the question posed by Drucker might still be open, or at least taken up in productive ways, particularly through dialogue and engagement rather than just interpretation.

---

24 F. Macdonald, 'Geopolitics and the "Vision Thing": Regarding Britain and America's First Nuclear Missile', *Transactions of the Institute of British Geographers* 31/1 (2006) 53–71, p. 55.

25 Drucker (note 10) p. 27.

26 W. J. T. Mitchell, 'What do Pictures "Really" Want?', *October* 77 (Summer 1996) 71–82.

27 D. Lisle, 'Benevolent Patriotism: Art, Dissent and *The American Effect*', *Security Dialogue* 38/2 (2007) 233–50, p. 234.

28 Drucker (note 10) p. 27.

In claiming to recognize the situated nature of knowledge it is necessary to be open about the fact that my own knowledge and experience is situated in particular ways. Though attempts at disclosure are never perfect (and potentially narcissistic), I hope these are clear to some extent at least. Of particular relevance is the fact that all reflections on the war on terror (particularly those emanating from, to use a reifying shorthand, the male, white, metropolitan West) take place in relation to orientalist geographies of power. With this in mind, it is worth recalling an early passage in *Orientalism* itself:

> In a quite constant way, Orientalism depends for its strategy on this flexible positional superiority, which puts the Westerner in a whole series of possible relationships with the Orient without ever losing him the relative upper hand.[29]

The issues this raises are not simple (not least to do with the apparent reifications involved), and they can be linked with other concerns about identity and subjectivity that have been raised about the project of critical geopolitics, in particular by feminist geographers.[30] While I do not claim that this provides a complete solution, my response is largely methodological, along lines suggested by Said's own idea of contrapuntal analysis: by engaging with and tracing between interventions emerging from different places within (and routes through) global geographies of power, it might be possible to discern something worthwhile about their working that could not be learned from any one position. This also follows Drucker's concern with 'points of connection, associations of reference, lines of accountability, responsibility, engagement'. As the artist Wafaa Bilal has written about his work the *Sorrow of Baghdad*,

> I can't take people physically to a war torn and embargo ravaged country, but I can do something analogous to it. If I can recreate this sensations that I have experienced, and that citizens of that country are now experiencing. If I can recreate the sensations that I have experienced, if I can surround the viewer with a sense of 'being there', this faraway region and the suffering found there becomes more real, more intelligible, to the viewer. I want the viewer to live 'the truth' for just a few minutes, to stir their emotions and to spark empathy with those who are oppressed.[31]

This passage expresses both the difficulty of communicating the experience of violence and oppression and the affective, embodied and performative registers involved. The conditioned reference to the truth also signals an awareness of its problematic nature, but not its complete irrelevance or exhaustion; rather, I would suggest, it calls for negotiated dialogue across geographical space and

29   E. Said, *Orientalism* (London: Penguin 2003) p. 7.
30   See the references in note 12 and the discussion in F. Macdonald (note 24).
31   <http://www.crudeoils.us/wafaa/html/bagdad.html>.

power differentials that is grounded in the embodied and emotional experiences of a wide variety of people and places. I therefore focus particularly on artists, artworks and curatorial strategies as they speak *between* and *across* some of the diverse geographies and subjectivities encompassed by the war on terror as much as *within* and *to* them.

Gemak and *Green Zone/Red Zone*

Green Zone/Red Zone, an exhibition and debate programme that ran from 20 October 2007 to 31 January 2008, was the inaugural project of Gemak. The rationale informing Gemak has been explained by its curator, Robert Kluijver, in ways that parallel closely Drucker's formulation of current epistemological and cultural predicaments, the goals of artists like Bilal and ideas of situated knowledge:

> The language of politics has been emptied of its meaning. This is a phenomenon that seems to intrigue, amuse or fire many artists. It also preoccupies me. Between the reality of our world and our rendering of it into language lies a gap that allows, almost demands individuals to come with their own interpretations. I can juxtapose or contrast these in various ways that accord more with my own experience of reality. I don't pretend this rendering is closer to the truth but I do hope that it is closer to the subjective reality experienced daily by so many non-Western people.[32]

Green Zone/Red Zone brought together the work of more than twenty artists from (in the senses of 'based in', and tracing 'roots to' and 'routes via') a range of countries (including Finland, USA, UK, China and the Netherlands) but with a particular orientation to Iraq (with artists from Baghdad, Falluja, Basra and Kirkuk as well as Iraqis based outside the country). The exhibition was composed of works produced in a wide range of media, including sculpture, painting, collage, photography, installation, cartography, video and poetry.

   Green Zone/Red Zone committed what Bourdieu terms the sin of vulgarity: many of its exhibits failed to meet prevailing standards for 'art'.[33] It included works by participants in the Open Shutters: Iraq project, which gave eight Iraqi women and one child digital cameras with which to document their lives, and short documentary videos by students from the Independent Film and Television College in Baghdad. Another element was the installation of the remains of a car largely destroyed in the March 2007 suicide bombing of the Mutanabi book market,

---

32 'Interview with Gemak's Robert Kluijver' (December 2005) at <http://www.labforculture.org/pl/Users/Site-Users/Site-Members/nat-muller/nat-muller/Interview-with-Gemak's-Robert-Kluijver>.

33   Cited in Luke, *Shows of Force* (note 6) p. 230.

a centre of intellectual life in Baghdad, which killed 38 people and injured more than 60. According to Kluijver, this exhibit elicited a wide variety of responses among visitors to the exhibition, including views from some critics that because it was not art, it did not belong in a gallery.[34] Within *Green Zone/Red Zone*, the car faced painted works by Hana Mal Allah addressing the destruction and division of Baghdad, her home city; the rust covered remains of the car resonating with the beige, brown, black and grey tones of the works on canvas.

Plate 14.1 Remains of car and works by Hana Mal Allah at
         Green Zone/Red Zone
Source: Photograph by Alan Ingram.

The exhibition also showed a series of maps produced by the Foundation for Achieving Seamless Territory (FAST), an architectural collective that aims to highlight the divisive consequences of top down architecture and planning.

---

34   Conversation with Robert Kluijver, 28 January 2008. In 2007 the British artist Jeremy Deller's proposal that a similarly damaged car from Iraq be placed on the vacant 'fourth plinth' in Trafalgar square in order to provoke discussion about the war was shortlisted for consideration by a commissioning panel sponsored by the Greater London Authority. See <http://www.london.gov.uk/fourthplinth/plinth/deller.jsp>.

The maps were produced by applying homeland security guidelines to different scenarios (revolt of the poor; ethnic revolt; flood; American invasion), but played out in the Hague itself rather than Baghdad or other non-Western cities.[35] The respatializations envisaged by FAST can be connected with Gemak's concern with the idea of the Hague as a place that is already spatialized as a multifaith, multicultural and economically unequal city.[36]

This briefest of overviews perhaps conveys something of the range of material gathered in *Green Zone/Red Zone*. What I would like to suggest is that this exhibition represented an unusually rich opportunity to make points of connection, to affirm associations of reference and lines of accountability, and to reflect on responsibility and engagement in the war on terror and emerging landscapes of security and insecurity. It did this by putting in play a series of connections and circulations around those landscapes, and foregrounding the interweaving of everyday life and geopolitics in ways that unsettle easy distinctions between them. In this way *Green Zone/Red Zone* presented a radical and contrapuntal remixing of the imaginative geographies that have structured and sustained the war on terror; a retwisting and refolding of our space/their space into a space where a kind of simultaneity is possible that is otherwise foreclosed by material geography and hegemonic performances of space.

Much of *Green Zone/Red Zone* can be said to have been concerned with urban geopolitics, a term elaborated by Stephen Graham to denote the sense in which 'the world's geopolitical struggles increasingly articulate around violent conflicts over very local, urban, strategic sites'.[37] In order to contextualize the exhibition and its components further it is worth exploring the geopolitical significance of the Hague as a particular kind of site within international relations.[38]

While Amsterdam is the main focus of The Netherlands art scene, the Hague is the seat of national politics. One referent for Gemak, then, is the reluctance of the Netherlands government to push for the application of international law in relation to the war on terror and its participation in aspects of it. However, the Hague is also host to a number of international institutions and a site where a range of instruments of international law, including those governing the conduct of war, have been concluded. In particular, the Hague is the site of the International Court of Justice (ICJ) (the principal judicial organ of the UN) and the International Criminal Court (ICC), the 'first permanent international judicial body capable of trying individuals for genocide, crimes against humanity and war crimes when national courts are unwilling to do so'.[39] The ICC has particular significance in light of the

35   See <http://www.seamlessterritory.org/The_Hague.html>.

36   'Interview with Gemak's Robert Kluijver' (note 32).

37   S. Graham, 'Introduction: Cities, Warfare and States of Emergency' in (ed.), *Cities, War and Terrorism: Towards an Urban Geopolitics* (Oxford: Blackwell) 1–25, p. 6.

38   I am indebted to Robert Kluijver and Rashad Selim for their insights into the urban and political geographies of the Hague.

39   Coalition for the International Criminal Court, at <http://www.iccnow.org/>.

fact that the US has consistently claimed that its privileged role as hyperpower and putative guarantor of global security, as well as the special nature of the US constitution in protecting US sovereignty, means that it cannot be subject to the Court, where it fears becoming the target of illegitimate and politically motivated proceedings. The US has also pressured other countries not to sign up to the Court and to agree waivers for US personnel. But it has gone further than this; in August 2002 President Bush signed into US law the American Servicemembers Protection Act, which authorizes *inter alia* the use of all necessary means (implying military force) to liberate citizens of the US or its allies being held by the ICC, giving rise to its nickname as the 'Hague Invasion Act'.[40] Though it might seem far removed from Guantánamo Bay, Diego Garcia or Bagram Airbase, the Hague has thus also been framed as a kind of exceptional space.

Rewriting Urban Geopolitics

This overview goes some way to setting up a discussion of a series of interventions by the leading Iraqi artist Rashad Selim that were developed at Gemak and shown at *Green Zone/Red Zone*.[41] Selim's recent work, which employs a variety of media and techniques, has been particularly concerned with the destruction of Baghdad, the city where he grew up, the effects of the invasion and war on its cultural life and heritage, and the responsibility of international actors for what has happened.[42]

His work at Gemak has been based around the *seb'a ayoun* (seven eyes) symbol, a protective charm given to newborn children across Iraq (see Plate 14.2, below). As Selim describes,

> Devoid of religious or political connotations it remains a modest yet potent symbol of well-wishing. I think this charm is based on a paradigm of non-random geometry found in nature: the gathering together of circles (spheres or rods) or the same diameter invariably gives us a hexagonal pattern rich in associations, universal in appeal.[43]

---

40   Human Rights Watch, 'US: 'Hague Invasion Act' Becomes Law' (August 2002) at <http://www.hrw.org/press/2002/08/aspa080302.htm>.

41   The interpretations advanced in this section emerged in large part from two extended conversations with the artist at the exhibition on 28 January 2008.

42   See biographical information at <http://www.incia.co.uk>, website of the International Network for Contemporary Iraqi Artists, and W. Cook, 'National Healing' (January 2004) at <http://www.newstatesman.com/200401190033>.

43   Text accompanying *A Symbol From Iraq, a Sign For Den Haag*, Rashad Selim 2007/8.

Selim also states, 'It symbolizes continuity, protection and natural law, a universal aspiration to peace and well-being.'[44] The symbol can be interpreted as carrying a number of geographical resonances across contemporary dividing lines. Its origins can be traced to early Mesopotamian civilizations, and one reading of the symbol links it to a cross section through a sheaf of wheat, referencing the beginnings of sedentary agriculture, urbanism and law in the region, events that connect present-day Iraq with Western civilization.[45] The symbol can be compared to the amulet commonly worn in Turkish society, and has been used by Selim in other works in combination with a variety of religious symbols. The references to hexagonal geometry also connect with modern numbering and mathematical systems that have roots spanning Arabic and Hellenic culture.[46] In one register, then, its use can be read as a reaffirmation of the intertwining of cultural heritages across the dividing lines constructed in orientalist imaginative geographies, as well as the deep roots of some of these heritages in places that have suffered immense cultural destruction.

The symbol has been used by Selim in a series of interventions. In the first, he made a stencil of the symbol to carry out a work of graffiti in the urban landscape. In this Selim, equipped with spraycans and accompanied by Rick Vogels, a photographer, went into the streets of the Hague. As he states, 'I felt the need to create a link, place *seb'a ayoun* among existing collections in the streets of Den Haag, explore and create statements reminding passers-by of Iraq.'[47] Selim sprayed the symbol together with the word 'IRAQ' in six locations, before being arrested along with Rick (it is possible they were spotted via the surveillance cameras that monitor the centre of the Hague). As Selim states:

> This was my first experience of overnight custody. The charge: violent damage
> to public property introduced new parallels with the invasion and destruction of
> Iraq. I agree laws must be respected and compensation for damage paid, within
> reason as at the local so on the international.[48]

Graffiti is a form with a long history and many variations in the present. In many US cities it is particularly associated with the territoriality of gangs, yet, as Selim

---

44   Rashad Selim, statement on poster for intervention at Dutch Parliament, February 2008 at <http://artxgeopolitics.wordpress.com>.

45   It is worth recalling that occupying American forces in Iraq have used the site of the ancient city of Babylon as a base and parking lot for heavy vehicles, causing huge damage.

46   A. Sen, 'What Clash of Civilizations? Why Religious Identity isn't Destiny' (March 2006) at <http://www.slate.com/id/2138731/>.

47   Text accompanying *Seb'a Ayoun Street "Graffiti" Art Intervention*, Rashad Selim, 2007/8.

48   Ibid.

notes, 'Graffiti has exploded across Baghdad and Iraq since the occupation'.[49] Indeed, the high concrete blast walls that have proliferated in Baghdad as part of US attempts to humanize a key aspect of the brutal redivision of the city have themselves begun to attract artworks, as have the similarly-constructed sections of the Israeli apartheid wall.[50] In *Green Zone/Red Zone*, the photographs taken by Rick were displayed along with explanatory text next to three photos of writing by US forces in Baghdad. The first depicts a US soldier crouching next to the wall of a building, upon the wall of which has been written in Arabic, 'this house contains terrorists'.[51] The second shows a US soldier who has turned to look at a building, in front of which stands a small child with hands over ears. On the wall of the building is a spray-painted red 'x', signifying 'this building has been cleared'. In the final photograph, a US soldier places his left hand on the right shoulder of a blindfolded Iraqi man, and with his right is using a marker pen to write a series of letters and numbers on the man's forehead as a means of identification. These three photos (which were gathered from the internet) are themselves an expression of the masculinist attempts of US forces to write and rewrite the urban landscape of Baghdad and the bodies of Iraqis themselves. Placed together with the photographs of Selim's graffiti intervention, they propose a contrapuntal geographical conversation about the relations of identity, law and power operating in each of the two spaces.

This first use of the *seb'a ayoun* plays into the second, which can also be read in terms of multidimensional, counter-geopolitical texts, relations and embodied performances. As well as a graffiti stencil, Selim created a road sign showing the *seb'a ayoun* in place of the usual symbol. His explanation:

> On a traffic sign the symbol engages a contemporary icon of law and order. It is in Mesopotamia where laws and world religions emerged to govern and bring order to urban society. It is in Den Haag where institutions have developed to uphold international law and human rights. Yet on Iraq these institutions remain silent.[52]

The sign was photographed at various locations around the Hague: outside sites of urban infrastructure, in a market in the city centre, and in front of a grocery shop catering to migrant communities, the public transport hub, the city administration, the ICJ and the headquarters of Shell. These six photos were formatted as circles

---

49   Ibid.

50   See <http://news.bbc.co.uk/1/hi/in_pictures/4230505.stm>, discussion by the artist Mark Vallen at <http://www.art-for-a-change.com/blog/labels/Artists%20and%20the%20Iraq%20war.html> and <http://www.incia.co.uk/4775.html>.

51   Translated by Rashad Selim in conversation with the author.

52   Rashad Selim, poster for intervention at Dutch Parliament, February 2008, at <http://artxgeopolitics.wordpress.com>.

and placed in the pattern of the *seb'a ayoun* on a poster printed in a limited edition of 50 to be distributed among the concerned or targeted institutions.

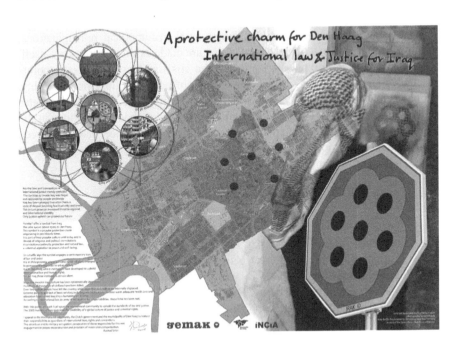

Plate 14.2 Poster for petition by Rashad Selim
Source: Rashad Selim for *Green Zone/Red Zone* © Gemak 2008.

For example, eight posters were distributed to the members of the foreign affairs commission of the Dutch parliament when they received Selim's petition on 19 February 2008. On the poster the *seb'a ayoun* design was accompanied by one of the scenario maps of the Hague produced by FAST, showing its respatialization as green zones and red zones under crisis conditions. Across the eastern edge of the city weaved the section of the River Tigris as it flows through Baghdad. On the right was a photo of Selim in the process of spraying a *seb'a ayoun* stencil during the graffiti intervention and the road sign itself. This process of layering and juxtaposition produced a complex geopolitical map of affiliation, connection and responsibility across diverse spaces and institutions.

On 26 January 2008 Selim took the road sign to the gates of the ICJ and, through a megaphone, read a petition. This included the following passage:

> With this public art work I call upon the international community to uphold the standards of law and justice … I appeal to the international community, the Dutch government and the municipality of Den Haag to honour their responsibilities

as guardians of international laws, rights and conventions. This entails an end to
military occupation, prosecution of those responsible for this war, engagement
in proper reconstruction and provision of reasonable compensation.[53]

I suggest that through a variety of contrapuntal geographical reimaginations,
enacted through processes of production that are made explicit in the work itself,
Selim's work offers a refamiliarization of the crisis in Iraq; it suggests 'points
of connection, associations of reference, lines of accountability, responsibility,
engagement' that stretch across the deepening borders of security. It creates
spaces within which to imagine different material and embodied locations in the
geopolitical present, and furthermore, suggests things that connect people between
them. It can be seen not only as a diagnosis of the present and a call for justice, but
maybe a kind of routemap as well.

Geographies of Responsibility, Technology and Power

These issues are taken up in different ways in the works by Peter Kennard and
Cat Picton-Phillipps exhibited as part of *Green Zone/Red Zone*. The biographies
of these artists trace rather different geographies from those of Selim and other
contributors, intersecting in other ways with the spatialities of the war on terror and
contemporary landscapes of security. They began their collaboration in response to
the Iraq war. Kennard is well known in the UK and internationally as a practitioner
of montage (he is often described as the greatest living exponent of the form) who
was radicalized by the Vietnam War and who has produced work for a wide range
of magazines and newspapers as well as social and political movements since
the 1970s, including a number of iconic images for the Campaign for Nuclear
Disarmament during the Cold War. Picton-Phillipps is a younger artist and printer
who has brought expertise in computer imaging as well as political imagination to
the collaboration. Their work has been made in a variety of media and exhibited in
a number of settings in the UK and elsewhere, and has focused particularly on the
roles of the US and UK, their senior political figures, corporations and the military,
the media, technology and ideologies of nationalism.

   Their work is undoubtedly interesting and provocative in its own right. But
what I focus on here is the way that it speaks with the work of others in the context
of *Green Zone/Red Zone*. I suggest that their dialogue across different locations
and positionalities in the war on terror multiplies and deepens their resonances in
suggestive ways, particularly regarding the question of responsibility.

   Kennard and Picton-Phillipps' participation in *Green Zone/Red Zone* is part
of an increasing engagement with artists and places outside the UK: in 2007 their
work was exhibited in Houston, Texas and in December 2007 they collaborated
with Palestinians in the West Bank, producing a large format work that was

---

53   Ibid.

placed on gates in the apartheid wall. Four of their works – *Control Room* (2006); *Surveillance* (2007); *Policy Papers* (2007); and *Presidential Seal* (2006) – were shown at *Green Zone/Red Zone*. These works (also large format) were produced by layering, tearing and exposing newspapers and photographic images and creating additional layers and representations with conventional art materials and oil, expressing in material form the multiple imbrications of media, power and commodities.

Plate 14.3 'Control Room'
Source: © Peter Kennard and Cat Picton-Phillipps, 2006.

Control Room, which faced from the gallery onto the street outside, depicts a centre of the kind from which much military action is now coordinated: a number of people sit at a bank of computer screens in a darkened room. But in place of the customary large screen is a wall whose fabric is a ripped and torn mix of dark areas and scratched-out washes of grey, with a variety of vague images. Instead of the promised transparent, panoptic vision of the battlespace, high technology here produces barely discernable fragments against a backdrop of chaos. In *Policy Papers,* Tony Blair and Gordon Brown are shown sitting together in a setting redolent of gendered state power, with wood panelled walls and flags. Yet they are surrounded by a tearing fabric of layered, decaying British (and Murdoch-owned) newspapers. In one part of the work these are torn back to expose Arabic newspapers otherwise hidden underneath. Scattered around the work are many smaller images of violence (including blindfolded, shackled and imprisoned male bodies) from the war on terror. *Presidential Seal*, which was shown opposite

Policy Papers, similarly presents a vast tableau (the work measures 3m by 6m) of layered and ripped newspapers, with numerous images of violence and destruction; of detainees in orange jumpsuits and torture. This forms a backdrop to a US presidential podium of the kind from which press conferences are given, yet here it is empty, awaiting the taking of responsibility for what is depicted. Finally, *Surveillance* shows in a similar way the increasing securitization of British urban space and the associated technologized surveillance and profiling of those who do not fit the norm, with key politicians, technologies, corporations and personnel depicted again in a large tableau. Visible behind and through this patchwork of images, on a scale approaching that of the work as a whole, is the face of Tony Blair.

Again, these works bring together, individually and collectively, a wide array of images and materialities that have been of central concern to geographers as they have tried to unpick landscapes of geopolitics and security, in ways that, I would suggest, can help to refamiliarize those landscapes and to make connections. *Green Zone/Red Zone* presented in a different way the networks of identity, technology, violence, narration and government through which geopolitics and security are currently constituted and of which much has been written. The construction of these mappings is both explicit in their form and explored in statements by the artists, as are their embodied, emotional and performative dimensions. As Kennard has stated (about *Obscenity*, an earlier exhibition at which *Control Room* and *Presidential Seal* were shown):

> It's all about trying to find ways to get a physical sense of the chaos and anarchy that's going on in Iraq. The work becomes much more physically-engaging and less intellectual ... It's more about the gut feeling we have about the situation and what this country has done.[54]

This both echoes Wafaa Bilal and makes explicit a concern with responsibility, as well as counterposing George W. Bush's oft-cited statement (both embodied and situated) that he is a 'gut player' rather than an intellectual. The production process also reflected the artists' dissatisfaction with a preceding series (*Award*), about which Picton-Phillipps has stated 'They became very beautiful, finished pieces, which was problematic for us. We wanted it to come off the wall more at people.'[55] Similarly, for Kennard, 'We wanted to make really big work ... The scale represents the scale of the destruction. We wanted it all to be much more physical than anything we had done before, and we hope that makes it easier for people to interact with it.'[56]

---

54   In A. Jones, 'The War on Error', *The Independent* (2 January 2007).
55   Ibid.
56   In 'Blairaq: Peter Kennard Interviewed About His New Exhibition', *Socialist Worker* 2057 (30 June 2007) at <http://www.socialistworker.co.uk/art.php?id=12201>.

Like other interventions in *Green Zone/Red Zone,* these works are situated knowledges. As Kennard has stated about *Blairaq*, an exhibition featuring *Policy Papers* and *Surveillance* at a gallery in east London that started in the week Blair left office:

> We never conceived that our work would be part of persuading people to change their minds ... Instead, we wanted to make art for the movement. We wanted people to be able to come and find their views echoed and validated by the work. To see in it the reasons why we marched and protested. Mainstream politicians and the media have conspired in creating the disaster in Iraq. We wanted to create something that makes people think that they are right to oppose the war, that everything we said about it was correct.[57]

Similarly, with *Policy Papers*, in which the room where Blair and Brown are sitting is decorated with wallpaper designed by the nineteenth century British socialist William Morris, Kennard 'wanted to say something about where the original utopian dreams of the [British] Labour Party have ended up'.[58] This perhaps is refamiliarization of a rather direct sort. It is true that a series of investigations have revealed the rather specific responsibility of Blair and his circle for British involvement in the invasion and occupation of Iraq.[59] But these works, if read geographically, have more complex implications than this suggests, first in view of their materiality: the use of newspapers (and in other related works oil and sand), official images, and depictions of technology call attention to much wider networks. Second, they allude (perhaps problematically) to the orientalist imaginative geographies abroad in the war on terror, notably an undifferentiated, distanced and schematic approach to the other, the display of their bodies, and a downplaying of their agency. In none of these works are the lives of people living in the violent zones of the war on terror shown in detail or in their own voices; they are shown as they are presented to most Westerners, through the prism of orientalism and corporate and state mediatization.

This is why *Green Zone/Red Zone* is so interesting for Kennard and Picton-Phillipps' work and for how we might think about art and the geopolitical. In *Green Zone/Red Zone*, these works speak to and involve each other and the viewer, but also work in relation to the rest of the exhibition and its location in the Hague. A work indicting Tony Blair for his role in the invasion of Iraq acquires extra valences by virtue of being located close to the ICJ and the ICC (as do works exploring the empire of oil when exhibited in Houston, the capital of the oil industry, in George W. Bush's adopted home state). Art, in a fairly obvious sense, takes place. At the same time, these works, which point to and foreground the

---

57   Ibid.

58   Ibid.

59   Notably J. Steele, 'Guys, I'm Afraid We Haven't Got a Clue ...', *guardian.co.uk* (21 January 2008) at <http://www.guardian.co.uk/politics/2008/jan/21/iraq.iraq>.

modalities of western power, spark against those (such as the films made by the Baghdad Film and Television College or the photos of the Open Shutters project that were also part of the exhibition) that speak of the ordinary continuities and problems of living, or, like those of Rashad Selim, challenge and collapse spatial divisions and distance and demand justice from powerholding institutions. They become part of a contrapuntal geographical exchange that constitutes a profound counter-mapping to the dominant geographies of the war on terror.

Conclusions

This chapter has argued that the compass of critical geopolitical scholarship can usefully be widened to take account of contemporary art practices, particularly those that are addressing the war on terror and contemporary landscapes of security. It has done so by bringing together conceptual tools from critical geopolitics and other forms of cultural criticism, and taking inspiration from convergences with artistic and curatorial strategies. There are numerous opportunities for further conceptualization and theorization, which could be pursued in a variety of ways. Similarly, the epistemological, ethical and political parallels and convergences the chapter identifies between strands of art and critical geopolitics open up the possibility for further conversations between them. To some extent such engagements are already evident in renewed interest in radical cartography, but they could be taken much further to extend wider debates about the relationship between art and politics.[60]

For example, one response to some of Kennard and Picton-Phillipps' work has been to argue that it insists on a simplistic realism, and not on the complexity of social totality.[61] This line of thinking can be questioned. First, as I have argued, the works are not as univocal as might be suggested, particularly if one seeks to read them geographically or in terms of the critique of orientalism. Second, to argue in this way is to assert a privileged political and epistemological position from which to dispense judgment on the meaning and value of aesthetic acts that I think is suspect (a 'peremptory verdict', as Luke argues, citing Bourdieu) and which pre-empts the possibilities opened up by ideas of situated knowledges.[62] It is not, as one critic suggests, that 'political art' is on a soapbox and needs to come down off it; still less that 'there are no political movements exposing and mobilizing against inequality and injustice'[63] (an absurd statement), but that it is possible to think

---

60   L. Mogel and A. Baghat (eds), *Atlas of Radical Cartography* (Los Angeles: Journal of Aesthetics and Protest Press 2007). See also E. O'Hara Slavick, 'Protesting Cartography: Places the United States Has Bombed', *Cultural Politics* 2/2 (2006) 245–254.

61   For example, D. Beech, 'Peter Kennard's *Decoration*', *Third Text* 19/1 (2005) 91–104.

62   Luke, *Shows of Force* (note 6) p. 230.

63   J. Appleton, 'Good Politics, Bad Art?', at <http://www.jjcharlesworth.com/the future/essays.html>.

more creatively about artworks in terms of relational geographical involvement in them and in what they address. I therefore propose a more situated and engaged approach that treats artworks not just as objects for critical reflection and judgement (a form of distinction, to be sure), but also sources of inspiration and provocation to further discussion, debate and intervention. This would align with a strategy of refamiliarization, which can also help to make explicit the ways in which people and places are now located within what some would call the battlespace.

Some critiques of contemporary 'political art' are also curiously static. The argument that Kennard and Picton-Phillipps were making 'solitary statements about the brutality of war' is surely unjustified, even before *Green Zone/Red Zone*.[64] But the curatorial strategy of placing their works together with those made by people from a wide variety of locations within and routes through global geographies of power opens up a still wider range of relational possibilities and events, without insisting on their meaning. These possibilities and events refer to the works themselves, the ways they relate to each other and the space of the exhibition, and the ways they reach out and address spaces both near and distant.

There are of course limits to such an enterprise. Not everyone can put on an exhibition or have their work exhibited, and urban spaces are not equally available to all. Where I have focused on synergies between the works, there are undoubtedly disjunctures and divisions too. Art may quite justifiably seem to be of little relevance to the reality of people living in zones of war, occupation and geopolitical tension. But the premise of this chapter has been that maybe these spheres are not quite so distant after all. I have suggested that the ways in which many artists and curators are engaging with the war on terror represents both a challenge and an invitation to geographers to reflect upon and engage differently in the reimagination of the geopolitical present, at the same time as they also seek to address it by more conventional scholarly means.

Acknowledgements

I would like to thank Klaus Dodds, Nick Megoran, Nadia Abu-Zahra and Jason Dittmer for their constructive criticisms and suggestions on this chapter; Robert Kluijver for corrections; and Robert Kluijver, Rashad Selim, Peter Kennard and Cat Picton-Phillipps for taking the time to talk at length about their work and for their permission to reproduce it. I am grateful to Hana Mal Allah for permission to include a photo containing images of her work and to Maysaloun Faraj for facilitating contacts and for suggestions.

---

64   Ibid.

Index